NIGHT SKY EXPLORER

YOUR GUIDE TO THE HEAVENS

ROBIN KERROD & TOM JACKSON

THUNDER BAY
P·R·E·S·S

Thunder Bay Press

An imprint of Printers Row Publishing Group
10350 Barnes Canyon Road, Suite 100, San Diego, CA 92121
www.thunderbaybooks.com • mail@thunderbaybooks.com

Correspondence regarding the content of this book should be sent to Thunder Bay Press, Editorial Department, at the above address. Author, illustration, and rights inquiries should be addressed to The Bright Press, Ovest House, 58 West Street, Brighton, BN1 2RA, United Kingdom.

Thunder Bay Press

Publisher: Peter Norton • Associate Publisher: Ana Parker
Editors: Traci Douglas, Angela Garcia
Senior Product Manager: Kathryn C. Dalby

Quarto Publishing plc

Publisher: Mark Searle
Associate Publisher: Emma Bastow
Creative Director: James Evans
Commissioning Editor: Sorrel Wood
Managing Editor: Isheeta Mustafi
Editor: Paul Sloman and Abbie Sharman
Designer: Paul Sloman | +SUBTRACT

ISBN: 978-1-64517-248-2

Printed in China

24 23 22 21 20   1 2 3 4 5

FSC
www.fsc.org
MIX
From responsible sources
FSC® C124807

Contents

Introduction

Every night of the year the whole world over, human eyes are raised to the sky to gaze and wonder at the starry heavens. The myriad of stars that shine down out of the velvety blackness of space still excite and fascinate as they did in the earliest times. Serious and systematic observations of the night sky began in the Middle East about 5,000 years ago, and in these observations lies the origin of astronomy, the scientific study of the heavens and heavenly bodies.

The early priest-astronomers who laid the foundations for what is now a very precise science had no idea what the heavens were like. Now, five millennia later, we think we know. The night sky provides us with a window to the wonders of the universe.

When we look at the stars, we are looking deep into that universe, traveling not only in space but also in time. The stars are so far away that the light we see has been traveling perhaps for thousands of years. We are seeing them as they were, not as they are.

Professional astronomers peer into space with their giant reflectors and use radio telescopes to tune in to the heavens and listen, as it were, to the symphonies of the celestial spheres. It is they who have solved many of the mysteries of our once incomprehensible universe, and they are beginning to work out what makes it tick. They are finding that in space, the fantastic is commonplace, the bizarre almost normal. And the more they find out, the more they discover there is to find out. Welcome to the universe!

◄ The Andromeda Spiral Galaxy, also known as M31, lies 2.5 million light-years from Earth.

Part One
Watching the Sky

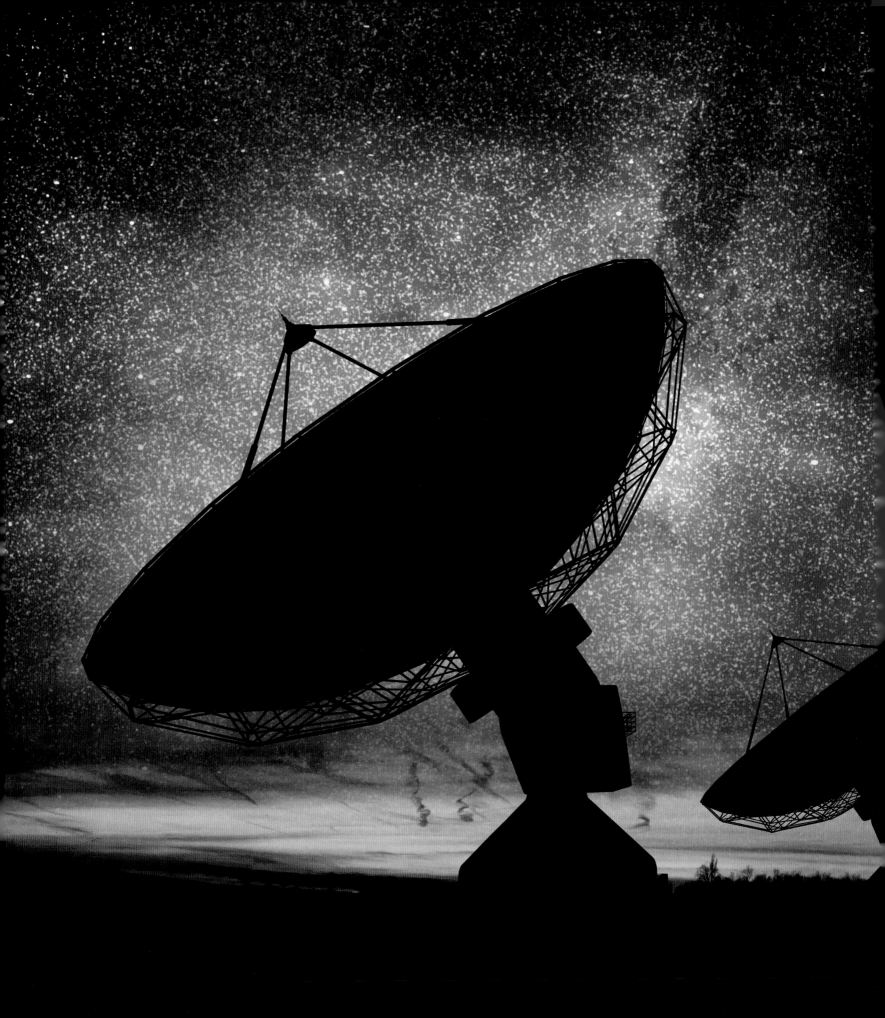

Studying the Heavens

At the simplest level, we need no special equipment to study the starry night sky. We can just look at it with our eyes and appreciate its splendor, just as our ancestors did long ago. But we can see a lot more with a little help. Binoculars or a small telescope will bring thousands more stars into sharp focus and show us mountains on the Moon, the changing shapes of the evening star, moons of Jupiter, and a host of other delights.

Professional astronomers use telescopes, too—giant ones. They also study the radio waves the heavenly bodies send out with huge radio telescopes. They send instruments into space on satellites and probes. With these powerful tools, they make the most astounding new discoveries about our solar system and universe.

◀ Radio telescopes point
toward a star-filled sky.

With the Naked Eye

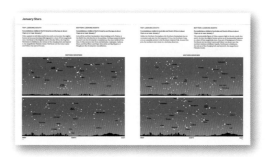

We can see an awful lot in the heavens just with our eyes. If we were very careful and very patient, we could count about 2,500 stars above the horizon at any one time. We notice that the stars are not all the same. Some are brighter than others. Most are white, but some are definitely colored.

Here and there we can spot what appear to be exceptionally bright stars, and, if we watch them over a number of nights, they move against the background of fainter stars. But these wandering heavenly bodies are not stars at all—they are planets. Most familiar is the bright evening star we often see shining in the west just after sunset. This is our nearest neighbor among the planets, Venus. Our nearest neighbor in space, of course, is the Moon. It is lovely to look at and ever changing, but it is so bright that it becomes a nuisance when you are stargazing!

Almost every time you go stargazing you will see streaks made by falling stars. These are meteors, specks of matter from outer space burning up in the atmosphere. The planets, the Moon, and meteors are all part of the Sun's family, the solar system.

BE PREPARED

If you intend to do any lengthy stargazing, it is worth making a few plans. First, you may need to wear warm clothing. It can be cool even on summer nights, while in winter, when the sky is usually at its clearest, it can be freezing.

To help you find your way around the sky, you will need a planisphere and a set of star maps. These are available as smartphone apps as well—and these have the advantage of being self-illuminating!

If you take a flashlight to read paper maps, it needs to have a red light so as not to affect your night vision.

▲ Check that star with your star map.

▶ A planisphere is indispensable.

Looking with Lenses

The eye is not a good "instrument" for viewing. It has only a small opening to let in the faint light from the stars. To gather more light, we use telescopes, which have a larger opening, or aperture.

Telescopes that gather light with lenses are called refractors, because they refract, or bend, the light passing through them. They have two lenses. The one in front, the objective, is the larger of the two. It forms an image that is then viewed by the eye through a second lens, called the eyepiece.

Amateur astronomers may still use refractors, but they cannot be built large enough for use in professional observatories.

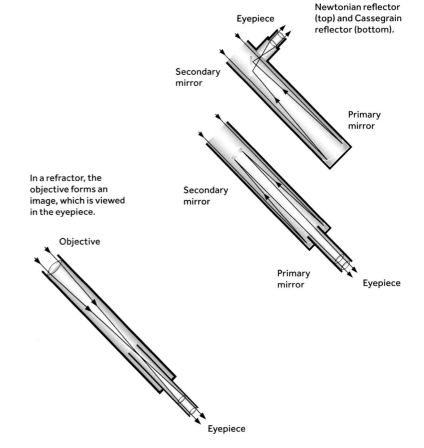

Newtonian reflector (top) and Cassegrain reflector (bottom).

In a refractor, the objective forms an image, which is viewed in the eyepiece.

MIRROR IMAGES

The telescopes that gather and focus starlight with mirrors are called reflectors. They gather the light in a dish-shaped, or concave, primary mirror. This mirror reflects the light onto a secondary mirror, which in turn reflects it into an eyepiece lens. In the most common type of reflector used by amateur astronomers, the eyepiece is located high up in the side of the body tube.

A reflector with a primary mirror of, say, about 8 inches (20 cm), is a useful size for amateurs. The giant reflectors used by professional astronomers have mirrors many times bigger. The two Keck reflectors at Mauna Kea Observatory in Hawaii have a 33-foot (10-m) primary mirror made up of 36 segments. The two Keck telescopes become even more powerful when they are used together to picture the night sky. The European Southern Observatory (ESO) in Chile has the VLT (Very Large Telescope), which uses four 26-foot (8-m) reflectors together. In 2024, the ESO will open the largest reflector ever—the 129-foot (39-m) ELT (Extremely Large Telescope).

▸ A 10-cm (4-inch) refractor like this Bresser Telescope (left) produces excellent images.

▸ 20-cm (8-inch) reflector telescopes, like this Orion SkyView Pro Equatorial Reflector Telescope (right), are an excellent choice for amateurs.

PEAK PERFORMANCE

Today's giant telescopes are sited in observatories high above sea level. Mauna Kea Observatory in Hawaii is one of the highest, at a height of more than 13,000 feet (4,000 m). High up, the air is thinner, contains much less dust, and is not so affected by swirling air currents. This cleaner, steadier air provides much better viewing conditions.

Astronomers seldom look through their telescopes these days. Instead, they use the telescopes as giant cameras. The longer the camera is left exposed in the telescope, the more light it stores. In this way, it can collect the light from very faint stars that would otherwise be invisible.

FOLLOW THAT STAR

As stargazers soon find out, stars do not stay in the same position in the night sky. They wheel overhead—rising in the east and setting in the west, just as the Sun does. If astronomers just pointed their telescopes at the heavens and left them for hours, all they would get would be blurred star trails. To avoid this, telescopes have to be driven around by a motor that matches the movements of the stars above.

SUNWATCH

Astronomers also use special reflecting telescopes to study the Sun. Usually, Solar telescopes take the form of tall towers with an observation room at the bottom. A mirror (called a heliostat) on top of the tower reflects light from the Sun down a shaft. Further mirrors reflect and focus the light onto a screen or table in the observation room.

▶ Top: The Kitt Peak National Observatory high up in the Arizona desert, also site of the McMath-Pierce solar telescope (bottom).

Filming the Skies

Capturing images of the heavens on film is called "astro-photography." As we have seen, professional astronomers record their observations in this way—and many amateurs do so, too. At the simplest level, there is nothing difficult about astrophotography. If you have a camera with a B (or time-exposure) setting, you can take pictures of the stars. You will also need a few other pieces of equipment—a cable shutter release and a tripod, preferably with a "pan-and-tilt" head that you can swivel around with a handle.

Mount the camera on the head of the tripod and screw in the cable shutter release. This prevents you from jostling the camera when you operate the shutter. Line up the camera with the part of the sky you want to photograph, then press the cable release to open the shutter. Clamp the release in the open position. You should leave the shutter open for some time, half an hour or more. Then release the cable and let the shutter close. Repeat the operation, pointing the camera in different directions. Vary the times of exposure. If you live in the Northern Hemisphere, point the camera toward the Pole Star (North Star) if you can find it. If you live in the Southern Hemisphere, point the camera toward the Southern Cross.

This produces pictures called "star trails," which cross the night sky in different directions. Your shots of the Pole Star or the Southern Cross should show the stars moving in circles.

CATCHING COMETS

Be sure to get out your camera when bright comets appear, which is not often. Again, mount your camera on a tripod with a cable shutter release. But this time use quite a short exposure, of about a minute. You should capture the comet well and find that this time the stars will be single points of light rather than trails.

▼ Pointing your camera at the Pole (or North) Star produces little arcs, with the near-stationary Pole Star in the center.

Seeing the Invisible

We see the stars because of the light they give out. Light is one way in which these bodies give off energy, but stars also give off energy as other radiation, including gamma rays, X-rays, ultraviolet, infrared, microwaves, and radio waves, which are invisible to our eyes. To gain more complete knowledge about the stars, we need to study their invisible radiation as well as their light.

However, there is a problem. Most of the invisible forms of radiation coming from the heavens are absorbed as they pass through Earth's atmosphere. Only the radio waves can reach us on the ground. Radio astronomers pick up heavenly radio signals with radio telescopes. Some take the form of sets of long wire antennae, but most are huge metal dishes. Radio astronomy has proved to be one of the most exciting fields of astronomy, giving us quite a different view of the universe.

Radio dishes are reflectors, which gather the signals and focus them onto an antenna. The faint signals are then fed to a receiver, where they are tuned and amplified (strengthened), as broadcast signals are in an ordinary radio set. The final signals are then transformed into visual images displayed in a variety of false colors. The radio images of heavenly bodies are often dramatically different from light pictures.

TELESCOPES IN SPACE

To study the other invisible radiation that permeates the heavens—the gamma rays, X-rays, ultraviolet rays, and so on—astronomers must go into space. They pack their telescopes and other instruments into satellites and launch them into orbit around the Earth, high above its absorbing atmosphere.

▲ The largest radio telescope, at Arecibo, Puerto Rico, as viewed from the observation deck.

▶ The Parkes radio telescope in New South Wales, Australia.

Hubble's Universe

The astronomy satellite that has captured the imagination like no other is NASA's Hubble Space Telescope (HST). It is a relatively conventional reflecting telescope with a light-gathering mirror only 95 inches (2.4 m) across. However, up in space, the HST has a perfectly clear view of the heavenly bodies and sends back images far superior to those obtained by much bigger instruments on the ground.

This was not always the case. After its launch from the space shuttle in April 1990, the HST was found to be flawed. The pictures it began sending back were little better than those obtained from Earth-based instruments. There had been a manufacturing fault that left the main mirror with an incorrect curve. In 1993, NASA astronauts fitted an arrangement of mirrors called COSTAR to correct the HST's faulty vision. Only then did HST live up to NASA's promise that it would open up a new window on the universe, and a series of upgrades—the last one being in 2009—have extended the life of the telescope greatly. One of the HST's most iconic images is the Ultra Deep Field, where the telescope spent three months of 2003 staring out into a tiny patch of seemingly empty space. The telescope saw hundreds of lights—entire galaxies—that are about 13 billion light-years away.

HST managed to look a little further in 2012, to 13.2 billion light-years, a few hundred million years after the Big Bang. However, to see further than this, back to the point in time when the very first stars and galaxies were forming, NASA will need a new telescope—the James Webb Space Telescope, which is scheduled for launch in 2020.

▼ The Hubble Space Telescope drifts 343 miles above the Earth's surface. The image on the left shows Hubble held in place by a Space Shuttle robotic arm during servicing.

▲ The Hubble Ultra Deep Field is the deepest visible light image ever made of the universe. It reveals some of the first galaxies to appear shortly after the Big Bang.

Looking Back in Time

To see further than the Hubble Space Telescope, nearer to the very edge of the visible universe, NASA has built the James Webb Space Telescope (JWST). Due for launch in 2020, this spacecraft's 21.3-foot-wide gold-plated mirror will be able to see objects 13.5 billion light-years away, and in so doing show us what the universe was like when the first stars were forming. However, to achieve that, the JWST will need to see infrared, or heat waves, not visible light.

Light and other radiation crosses the universe at a fixed speed—186,000 miles (300,000 km) per second. This is fast, but far from instantaneous. It takes eight minutes for sunlight to reach Earth, for example. The Sun is eight light-minutes away, which means we see the Sun as it appeared eight minutes ago. In the same way, light from objects 13.5 billion light-years away has taken 13.5 billion years to reach us—and shows objects from the early universe. However, in the billions of years that light has spent traveling toward us, the space around it has been expanding. As the light moves through expanding space, its wavelengths are stretched—turning the most ancient light sources into invisible infrared.

To pick up the faint glow of distant heat, the JWST is being put in a very cold and dark orbit called L2. L2 is a point 1.5 million km (930,000 miles) beyond Earth's orbit, where JWST will always be shielded from the Sun's heat by Earth's shadow.

The telescope is also shrouded by a tennis-court-sized heat shield, which ensures heat sensors are always cold enough, at least −223°C (−370°F), to pick up infrared from deep space. NASA says that the telescope will be so sensitive it could detect the heat of a bumblebee on the Moon. Astronomers are drawing up their plans for this latest instrument.

▶ Top: The James Webb Space Telescope's mirror is made from 18 hexagons that were folded away for launch.

▶ Middle: It will also be able to see into dark dust clouds in which stars are forming. No telescope has been able to see this clearly yet.

▼ Bottom: Webb will orbit the sun 1 million miles (1.5 million km) away from the Earth at what is called the second Lagrange point or L2 (graphics not to scale).

▶ Facing page: The telescope's hexagonal primary mirror at NASA's Goddard Space Flight Center.

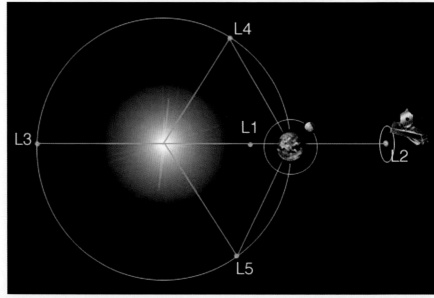

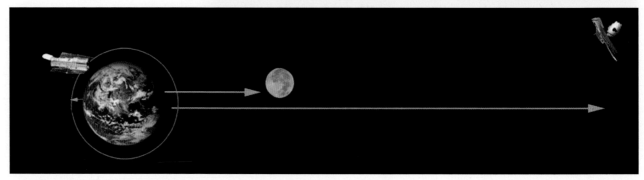

Hubble (345 miles)　　　　　　　　Moon (238,900 miles)　　　　　　　　Webb (1 million miles)

Probing the Planets

Just as the Hubble Space Telescope has opened up a new window on the universe, so space probes have opened up new windows on the planets and other bodies in the solar system. The Mariner 4 probe sent back the first close-up picture of Mars in 1965, showing a bare landscape. Later Mariners revealed a planet of contrasting landscapes, with deserts, volcanoes, and great canyons.

Mariner 10 reported back from Mercury (1974), revealing a cratered wilderness. Pioneer 10 journeyed through the asteroid belt to visit Jupiter (1973) and two identical Viking probes released landing craft that set down on Mars in 1976, taking pictures of the rust-red landscape.

DEEP-SPACE VOYAGERS

Another pair of robot craft, *Voyagers 1* and 2, set off from Earth (1977) to visit Jupiter and Saturn in turn (1979–81). *Voyager 2* sped on to fly past Uranus (1986) and Neptune (1989). Both *Voyagers* are still in operation, reporting back about conditions at the very edge of the solar system. By the time their power supplies run out, in 2025, they should be traveling in interstellar space.

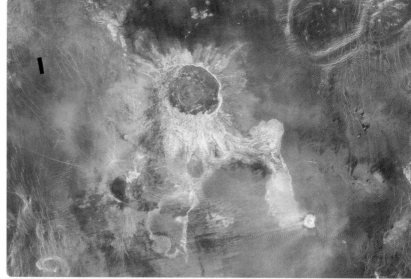

LOCAL EXPLORERS

Much nearer home, *Magellan* (1990–94) has used radar to peer through the thick cloud of Venus, revealing a landscape of dead volcanoes and lava plains. *Galileo* (1995–2003) and *Cassini* (2004–2017) have toured the Jupiter and Saturn systems respectively—with the latter dropping a lander onto Saturn's moon Titan (2004) to reveal a world covered in lakes of gasoline. In 2011, while orbiting Mercury, *Messenger* began four years of observation. However, Mars has received the most attention, with the rovers *Spirit* and *Opportunity* arriving in 2004 and *Curiosity* landing in 2012. *Spirit* became bogged down in sand in 2010, but the other two continue to explore the red planet.

▲ Top: *Voyager 1* captures the amazing ring system of Saturn in 1980 after visiting Jupiter.

▶ Bottom: *Magellan's* radar eyes spot vast lava flows all over Venus.

▶ Facing page: *Galileo* captures Enceladus, covered in a thick crust of ice with an ocean beneath containing twice as much liquid water as Earth.

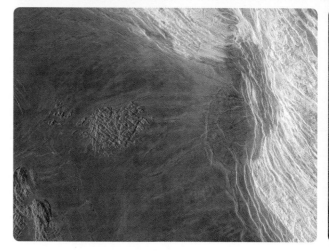

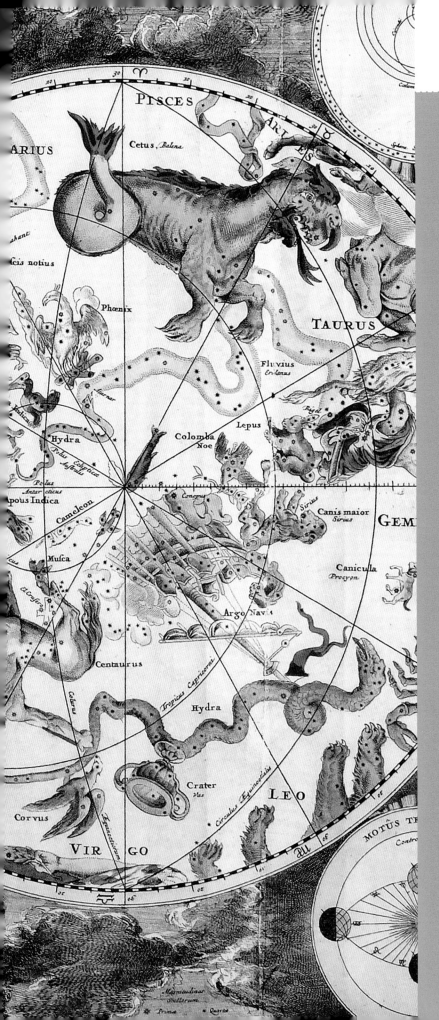

Patterns in the Sky

Stars in the thousands shine down on us out of the blackness of the night sky—the blackness that is space. They seem to be scattered haphazardly in every direction we look. At first sight, therefore, the heavens appear to be somewhat confusing. But stargaze for a while, and your eyes will soon begin to discern some order in the apparent chaos of stars.

You will notice that some stars are brighter than others and seem to make recognizable patterns. If you look in the same part of the heavens at about the same time several nights in a row, you will find the same star patterns there.

We call these recognizable patterns of bright stars the "constellations." They provide an invaluable means of guiding us around the night sky, just as they did our distant ancestors, who saw much the same skies as we do today. It was they who named most of the constellations—Orion, Scorpius, Cetus, Cassiopeia, Centaurus—and identified them with the heroic, the beautiful, the fantastic, and the monstrous characters that were featured in their epic, and often racy, myths and legends.

◄ A colorful star map from the early eighteenth century showing some of the constellations of the Northern Celestial Hemisphere.

The Constellations

Astronomers recognize 88 constellations, from Andromeda to Vulpecula. Many of them have been with us since the dawn of astronomy in Babylonia and Egypt more than 5,000 years ago. Astronomers of later civilizations added new ones as time went by, most recently in the 1700s.

The Greeks recognized 48 constellations and gave them names that reflected what they thought the constellation patterns looked like. Here was a scorpion, with its curved tail ready to strike; there a crouching lion; here an archer; there twin boys; and there again a chained lady. Some of the constellations do passably resemble the figures they are meant to represent, including the scorpion and the lion. But in most cases there is little resemblence. The ancient Greeks obviously had very vivid imaginations!

THE CONSTANT CONSTELLATIONS

The constellations never seem to change. They will look the same throughout your lifetime and for many millennia to come. All the stars in the constellations appear to be fixed forever in place and travel together through space. But this is far from the truth.

With few exceptions, the stars in the constellations are different distances away. We only see them as a group because they happen to lie in the same direction in space. And the stars in the constellations are not fixed in place—they are all traveling in different directions. We can't see them move perceptibly over time because they are too far away. But gradually, over many tens of thousands of years, they will change their position in the heavens. Then the scorpion and the lion, the archer and the chained lady, the dogs and the bears that grace today's skies will be history.

▲ Above: Ancient astronomers pictured Orion as a mighty hunter, with his right arm raised, ready to strike a blow with a hefty club.

◀ Left: One of the finest constellations, the unmistakable Orion.

◀ Facing page: This stunning image shows the Orion Nebula (M42), a huge cosmic cloud of interstellar gas located in the Sword of Orion, and situated to the south of the trio of bright stars forming Orion's Belt.

The Dome of the Heavens

Our ancient ancestors pictured the heavens as a vast sphere surrounding the Earth, and it is easy to understand why. From anywhere on Earth, the night sky envelops you like a great dark dome, studded on the inside with stars. Since the sky appears as a dome everywhere, the heavens must therefore be a sphere around the Earth—a great celestial sphere. The sphere, of course, spins around, carrying the stars with it—we see this happen every night. It spins from east to west, for that is the direction in which the stars move at night, and the Sun moves by day.

But the spinning celestial sphere is an illusion. The heavens in reality stretch out to unfathomable reaches of space, where distances are measured in light-years—units of millions of millions of miles. And the east–west movement of the stars across the sky at night, and of the Sun by day, happens because Earth is spinning around in space from west to east.

An illusion it may be, but the celestial sphere is far from an outdated concept. It accurately reflects our view of the heavens and provides astronomers with a means of pinpointing the stars. Making use of the geometry of a sphere, they locate a star using a system of coordinates—grid references—similar to the lines of latitude and meridians of longitude system geographers use to pinpoint places on the spherical Earth. In addition, the celestial sphere has a line equivalent to Earth's equator, an imaginary line midway between the North and South Poles. The celestial equator is a line midway between the North and South Celestial Poles. Stars travel parallel with the celestial equator, and the line divides the celestial sphere into the Northern and Southern Celestial Hemispheres.

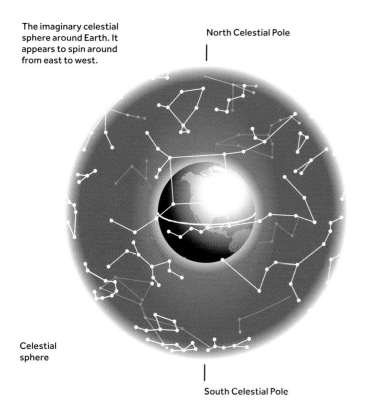

The imaginary celestial sphere around Earth. It appears to spin around from east to west.

North Celestial Pole

Celestial sphere

South Celestial Pole

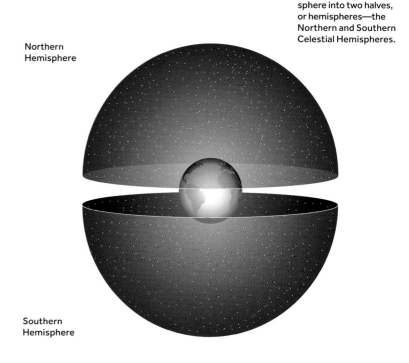

For convenience, we divide the celestial sphere into two halves, or hemispheres—the Northern and Southern Celestial Hemispheres.

Northern Hemisphere

Southern Hemisphere

Here the constellations are listed in alphabetical order, with maps to show the position of the major constellations in the Northern and Southern Celestial Hemispheres.

1	Andromeda	13	Canes Venatici (Hunting Dogs)	25	Coma Berenices (Berenice's Hair)	35	Equuleus (Colt)
2	Antlia (Air Pump)	14	Canis Major (Great Dog)	26	Corona Australis	36	Eridanus (River)
3	Apus (Bird of Paradise)	15	Canis Minor (Little Dog)		(Southern Crown)	37	Fornax (Furnace)
4	Aquarius (Water Bearer)	16	Capricornus (Sea Goat)	27	Corona Borealis	38	Gemini (Twins)
5	Aquila (Eagle)	17	Carina (Keel)		(Northern Crown)	39	Grus (Crane)
6	Ara (Altar)	18	Cassiopeia	28	Corvus (Crow)	40	Hercules
7	Aries (Ram)	19	Centaurus (Centaur)	29	Crater (Cup)	41	Horologium (Clock)
8	Auriga (Charioteer)	20	Cepheus	30	Crux (Southern Cross)	42	Hydra (Water Snake)
9	Boötes (Herdsman)	21	Cetus (Whale)	31	Cygnus (Swan)	43	Hydrus (Little Water Snake)
10	Caelum (Chisel)	22	Chameleon	32	Delphinus (Dolphin)	44	Indus (Indian)
11	Camelopardalis (Giraffe)	23	Circinus (Compasses)	33	Dorado (Swordfish)		
12	Cancer (Crab)	24	Columba (Dove)	34	Draco (Dragon)		

Viewing the Constellations

Exactly which constellations you will see in the heavens when you go stargazing depends on many factors. One is the time of observation. Because the Earth is spinning around on its axis, the stars move across the sky. So all through the night the stars are rising, moving across the heavens, and then setting. Constellations that were visible in the early evening may have disappeared before the Sun rises the next morning.

Because the Earth is a round ball, people living at different latitudes—distances away from the Equator—will see different night skies. Observers in Canada, for example, will be familiar with far northern star groups such as the Big Dipper, or Plough (Ursa Major), and Cassiopeia. But they will never be able to see far southern constellations such as the Southern Cross or the brilliant star Canopus. Conversely, South Australians will find the Southern Cross a familiar constellation, but will never glimpse the Big Dipper (Plough) or see the Pole or North Star, Polaris.

◀ Facing page: The night sky is dense with stars and nebulae.

▼ Ursa Major and Ursa Minor: key constellations for finding your way around the heavens.

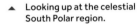
▲ Looking up at the celestial South Polar region.

◀ The night sky viewed from equatorial latitude.

▶ How the stars appear to move when viewed from mid-latitudes in the Northern Hemisphere.

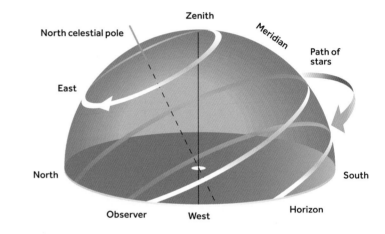

Zenith

North celestial pole

Meridian

East

Path of stars

North

South

Observer

West

Horizon

HOW THE HEAVENS MOVE

Where you happen to live on Earth not only determines which stars you see, it affects the way you see them move across the sky. If you live at mid-latitude in the Northern Hemisphere—for example, in the United States or Western Europe—the stars move as shown in fig. A (right). They rise above the eastern horizon, arc across the sky, and eventually set below the western horizon. If you look north, the stars appear to circle counterclockwise around the North celestial pole and nearby Pole Star. If you look south, the stars travel in an arc clockwise from east to west, culminating, or reaching their highest point, in the sky due south (fig. B).

But, in other locations, you would see the stars move in a different way. At the North and South Poles, for example, you would see the stars travel parallel with the horizon. At the Equator, you would see the stars rise and set vertically. The Equator, incidentally, is one of the best locations for stargazing. During the year, an observer there will be able to see all the constellations in the heavens at some time, a luxury denied to observers at other latitudes.

A Looking south, they arc across the sky from east to west.

B Looking north, the stars circle around the Pole Star.

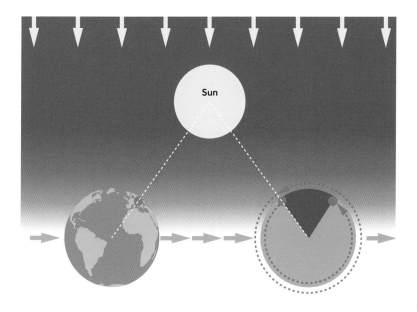

▼ Star-time days are slightly shorter than normal days.

THE SUN'S PATH

Although we know that Earth circles around the Sun each year, it appears from our Earthly viewpoint that the Sun travels in a great circle around the heavens, around the celestial sphere. The path it takes is called the ecliptic.

The Moon and the planets travel through the heavens within an imaginary band around the ecliptic. This happens because they travel through space in much the same plane as Earth. The imaginary band is called the Zodiac, and the 12 star groups it passes through are known as the constellations of the Zodiac. The word "Zodiac" means something like "circle of animals," because most of the constellations have animal names, such as Taurus (the Bull) and Aries (the Ram).

In astrology, people born under certain signs are supposed to possess certain traits, and the positions of the planets within the Zodiac are supposed to influence people's lives.

Astronomy reveals fundamental flaws in this concept. The Sun actually passes through 13 constellations on its annual journey (the extra constellation is Ophiuchus). And the Zodiac used by astrologers is based on the path of the Sun through the heavens in Roman times. Because of precession, a wobbling of Earth's axis, the Sun currently passes through the constellations roughly a month earlier, so all the star signs are out of synch.

STAR TIME

The stars whirl around the night sky because Earth rotates on its axis, spinning around once in 24 hours. To be precise, the spinning Earth returns to the same position in *relation to the Sun* every 24 hours. This is the time period we call the solar day. However, as Earth is spinning around, it is also moving along its orbit around the Sun. So, after 24 hours, it has had to spin around slightly more than once in space in order to end up in the same position in relation to the Sun (see diagram).

Earth actually spins around once in space every 23 hours 56 minutes. This is Earth's true period of rotation and forms the basis of "sidereal" time—time measured in relation to the stars, not the Sun. What this amounts to in the heavens is that any particular star rises, culminates and sets four minutes earlier every night, according to normal (solar) time. But if we use star time—sidereal time—things get much simpler. The whirling heavens keep pace with this time. A star will always rise, culminate, and set at the same time, and it will be in the same position on the celestial sphere at the same sidereal time, always. As a result, sidereal time is used to define celestial longitude (right ascension). Sidereal time is also discussed on page 52.

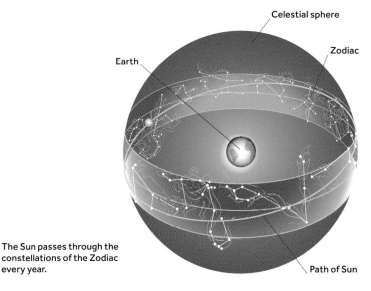

The Sun passes through the constellations of the Zodiac every year.

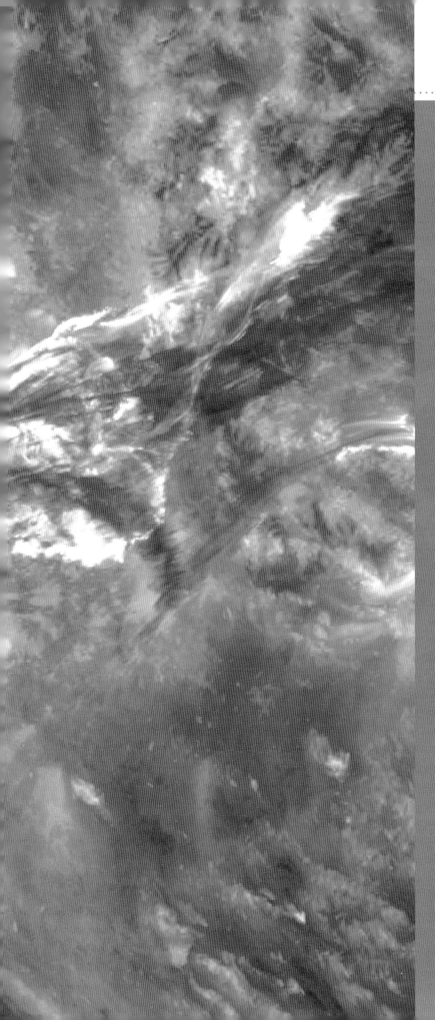

Family of the Sun

Four thousand, six hundred million years ago there was nothing in our tiny corner of the universe but a vast cloud of dark, cold gas, and dust. Then supernova shock waves rippled through the cloud, triggering an unstoppable process that eventually led to the birth of the Sun and the family of bodies that travel with it through space.

The senior members of the Sun's family, or solar system, are the planets, bodies that travel in circles around the Sun. Earth is one of these bodies. We can see five more wandering across the night sky like errant stars, and two others in our telescopes. That makes eight in all.

Most of the planets have companions, or "satellites." Earth has one, the Moon, and this has led to the satellites of other planets being called moons too. Many other bodies populate interplanetary space. The largest ones are "dwarf planets," such as Pluto and Ceres, and asteroids, but these are all too faint and lie too far away to see. We can, however, see the comets, which at their most spectacular can outshine the stars and stretch halfway across the heavens. The smallest members of the Sun's family are the "meteoroids," little bits of rock and metal that rain down on Earth all the time as the fiery streaks we call meteors.

◀ A coronal mass ejection from the Sun just missed Earth in 2012.

33

▲ A montage of eight planets seen from Mercury, with some of the larger moons.

The Solar System

In order of distance from the Sun, the planets are Mercury, Venus, Earth, Mars, Jupiter, Saturn, Uranus, and Neptune. The illustration below shows their orbits around the Sun. The inner planets are relatively close together, the outer ones very far apart.

It is difficult to appreciate just how big the solar system is. The distance from one end of Neptune's orbit to the other is nearly 5.5 billion miles (9 billion km). Pluto and most of the dwarf planets exist in a belt of icy bodies beyond Neptune, while most comets have fallen in toward the Sun from a "cloud" of similar bodies at the edge of the system that is light-years away, nearly halfway to the nearest stars.

HOLDING TOGETHER

The Sun is massive, more than 300,000 times the mass of Earth and 750 times the mass of all the rest of the bodies in the solar system put together. Being this big means that it has a very powerful pull, or gravity, that still exerts influence billions of miles away. The Sun's gravity alone is enough to hold the solar system together.

SUNSHINE

The Sun is the only body in the solar system to produce and give off light of its own. We see the other planets, their moons, and the myriad other bodies in interplanetary space only because they reflect the light from the Sun.

▼ Earth is the largest of the four rocky planets in the heart of the solar system. The four outer planets and Pluto, the largest dwarf planet, lie very far apart, separated by vast distances.

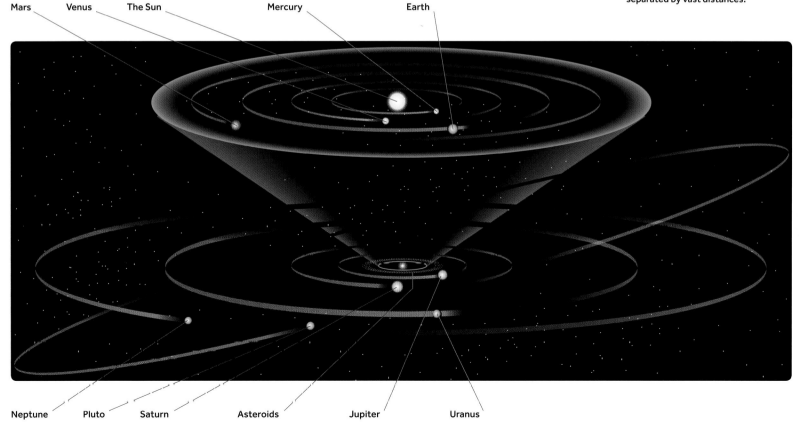

Mars   Venus   The Sun   Mercury   Earth

Neptune   Pluto   Saturn   Asteroids   Jupiter   Uranus

The Sun, our Neighborhood Star

A favorite trick question in quizzes is: which is the nearest star? Many answer: Proxima Centauri. But they are wrong, of course. The nearest star travels through our skies every day—it is the Sun.

The Sun is a typical star, a searing mass of incandescent gas, which pumps out fantastic amounts of energy into space as light, heat, and other forms of radiation. Whereas the light from the next nearest star takes more than four years to reach us, light from the Sun brightens our world in about eight minutes.

Like all ordinary stars, the Sun is made up mainly of hydrogen and helium. But it also contains traces of as many as 70 of the 90 or so elements we find on Earth.

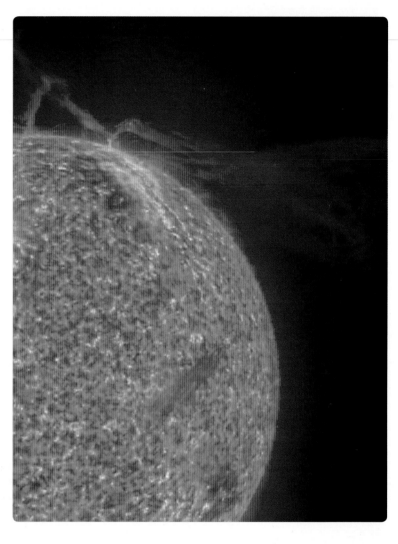

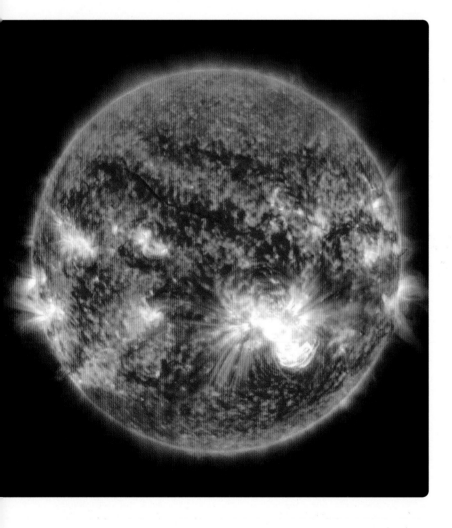

IN THE SOLAR FURNACE

The energy that keeps the Sun shining is produced in its center, or core. There, under great pressure and at temperatures of millions of degrees, nuclear reactions take place. The nuclei of hydrogen atoms fuse together to form helium atoms. The fantastic amounts of energy generated in this nuclear fusion process are equivalent to millions of hydrogen bombs going off every second.

The energy produced in the core radiates very slowly through the dense plasma, taking 100,000 years to get near the surface.

From the surface, known as the "photosphere," the energy radiates into space. A constant stream of electrified particles also escapes from the surface and flows out into space, forming the so-called "solar wind." The temperature on the surface of the Sun is about 9,900°F (5,500°C). This makes the Sun a medium-hot star, which gives out a yellow light. Astronomers class the Sun as a "yellow dwarf," of spectral group G2.

▲ The extraordinary power of the Sun is visible to us on Earth every day.

◀ Facing page: Epic solar prominences erupt from the Sun.

THE SUN'S ATMOSPHERE

Above the photosphere, the Sun has an atmosphere of thinner gases. First comes the "chromosphere," so called because of its generally pink color. Further out is the "corona," the outer atmosphere. It extends for millions of miles, thinning all the while, until it merges into space. The chromosphere and corona cannot usually be seen because the photosphere is so bright. Only during a total eclipse of the Sun do they become visible (see page 38).

THE STORMY SUN

The surface of the Sun is a seething, boiling mass of red-hot gas. In photos, it has a granular appearance, with each granule representing a pocket of rising gas. From time to time darker spots appear. These sunspots come and go over a period of about 11 years, known as the "sunspot cycle." Sunspots appear dark because they are about 1,800°F (1,000°C) cooler than the rest of the surface.

Brighter-than-average spots or patches on the surface are huge explosions known as "solar flares." They release a gale of solar wind, and this causes beautiful auroras on Earth. The most spectacular features of the Sun are its prominences. These are great fountains of flaming gas that shoot high into the chromosphere, following the invisible loops of the Sun's powerful magnetic field.

Eclipsing the Sun

The Sun, Earth, and the Moon move in much the same plane (flat sheet) in space, so there are times when they line up exactly. This happens about two to three times a year. Sometimes the Moon moves between the the Sun and Earth, and at other times Earth moves between the Sun and the Moon.

By coincidence, the Moon appears almost exactly the same size as the Sun in our sky. It is 400 times smaller than the Sun, but 400 times closer. So, when the Moon moves between the Sun and the Earth, we see it cover the Sun and blot out its light. This is a "solar eclipse."

THE MOON IN ECLIPSE

When Earth moves between the Sun and the Moon, it casts its shadow on the Moon, creating a lunar eclipse. The Moon can stay in eclipse for more than two hours. But it never completely disappears because of light refracted around Earth by the atmosphere. Often it takes on a pink or orange hue.

▼ ESA's PROBA-2 View of a solar eclipse, March 2015.

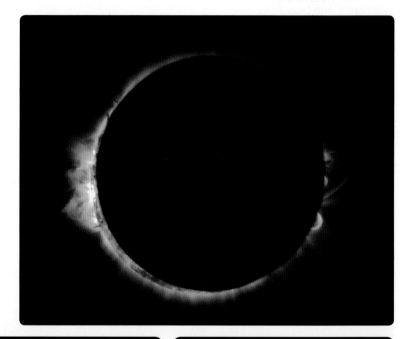

A TOTAL EXPERIENCE

Total solar eclipses are among the most spectacular sights on Earth. As the Moon edges across the Sun, the light gradually begins to fade. Then it is twilight, with only a sliver of sunlight peeping out. In seconds, darkness falls as the eclipse becomes total. Day turns suddenly into night and the air chills.

The pearly white halo of the Sun's outer atmosphere, the corona, becomes visible during totality. But the spectacle is over all too soon—eclipses never last for more than a few minutes. The edge of the Moon brightens, and a bead of sunlight emerges, creating the effect of a sparkling diamond ring. In seconds, daylight returns.

Total eclipses can be seen only along a "path of totality" created by the shadow of the Moon racing along at a speed of more than 1,000 miles per hour (1,600 km per hour). Because the Moon casts only a small shadow in space, this path is never more than about 170 miles (270 km) wide.

▶ Bottom left: The Moon lit by red light shining through the Earth's atmosphere in a solar eclipse.

▶ Bottom right: The Moon takes a "bite" out of the Sun during a partial solar eclipse.

The Moon, Queen of the Night

The Moon is our nearest companion in space, about a hundred times nearer than the closest planet, Venus. It is quite a small body, with a diameter about a quarter of that of Earth and a sixth as massive. It has low gravity, and as a result it has not been able to hold on to any atmosphere.

During the lunar day, which is two Earth weeks long, temperatures can climb to more than 250°F (120°C). And during the equally long lunar night they can plummet to −240°F (−150°C).

▼ A "super blue blood" Moon near California's NASA Armstrong Flight Research Center.

CLOUDS AMONG THE STARS

Like all moons, ours has two motions in space. One, it spins around on its axis like a top. And, two, it circles around its parent planet. The Moon spins around on its axis once every 27 1/3 days. And it also takes the same time to travel once around Earth. The result of these identical periods is that the Moon always presents the same face toward us. We say that the Moon has a "captured rotation," and it is far from unique in this respect. Mercury is locked to the Sun in the same way.

The Changing Moon

The Moon shines only because it reflects sunlight. As the Moon travels around Earth, we see different amounts of it lit up. Sometimes we see it only as a thin sliver, while at others we see it full-circle. We call these changes "lunar phases."

The Moon goes through its phases in 29 ½ days, a period that is the origin of our month. Today's calendar months have been altered to bring the calendar in line with Earth's year.

FROM CRESCENT TO FULL

The first phase of the Moon is the "new moon," in which the Moon is actually invisible! This is because the Sun is lighting up the Moon's far side. After a day or so, we see a sliver of light appear around the edge of the Moon as a slim crescent. This grows larger as the days go by, and, a week after the new moon, the Moon has become a half-circle, or the first quarter. After another week, the Moon has become a full circle, and we have the full-moon phase. A growing Moon is said to be waxing. Over the next two weeks the Moon wanes, or shrinks back into an invisible disk.

TIME AND TIDE

The gravity of the Moon is strong enough to tug at the waters of the oceans and pull them away from the center of Earth. This creates a high tide directly under where the Moon happens to be as Earth spins around. Low tides on either side are caused by the water being tugged away, and there is also a high tide on the opposite side of Earth due to the inertia of the water there.

First Quarter

Waxing Gibbous

Waxing Crescent

Full

New

Waning Gibbous

Waning Crescent

Third Quarter

▶ The phases of the Moon.

Moonwatching

The main seas and prominent craters of the Moon. Apollo landing sites are also marked.

North

SEA OF COLD

Aristoteles

Pluto

Alps

Eudoxus

Bay of Rainbows

Caucasus Mtns

Jura Mtns

Aristillus

SEA OF SHOWERS

SEA OF SERENITY

Cleomedes

Archimedes

+ Apollo 15

Haemus Mtns

+ Apollo 17

Apennines

Eratosthenes

SEA OF CRISES

SEA OF VAPORS

Manilius

Plinius

SEA OF TRANQUILITY

OCEAN OF STORMS

Kepler

Copernicus

West

Reinhold

+ Apollo 11

SEA OF FERTILITY

East

Grimaldi

+ Apollo 12

+ Apollo 14

Hipparchus

+ Apollo 16

Ptolemaeus

Theophilus

Langrenus

Albategnius

SEA OF NECTAR

Alphonsus

Gassendi

Arzachel

Vendelinus

SEA OF MOISTURE

SEA OF CLOUDS

Purbach

Fracastorius

Petavius

Pitatus

Piccolomini

Walter

Stofler

Maurolycus

Schickard

Tycho

Janceon

Maginus

Longomontanus

Clavius

Bailly

South

Lunar Landscapes

The Moon is so close that we can make out two different kinds of features with the naked eye—dark areas and light areas. Early astronomers called them maria (Latin for "seas") and terrae (Latin for "land"). As telescopes came into use, it became apparent that the dark areas were vast plains, with not a drop of water in sight. Nevertheless, we still call the dark regions maria or seas.

The seas are the biggest features on the Moon. The Ocean of Storms (Oceanus Procellarum) alone is as big as the Mediterranean Sea on Earth. It is a great sprawling expanse of ancient lava flows. Like the other seas, it has many fewer craters than the light-colored highland regions. This shows that the seas are younger. The highlands are thought to be part of the Moon's original crust. Many of the seas are circular in shape and were formed when lava flooded into existing huge craters some 3 billion years ago. Lofty mountain ranges ring many of the craters, soaring to heights of more than 16,000 feet (5,000 meters).

▼ An oblique view of the Sea of Tranquility.

▶ A striking view of the Moon with Earth in the background, seen from an Apollo mission in orbit.

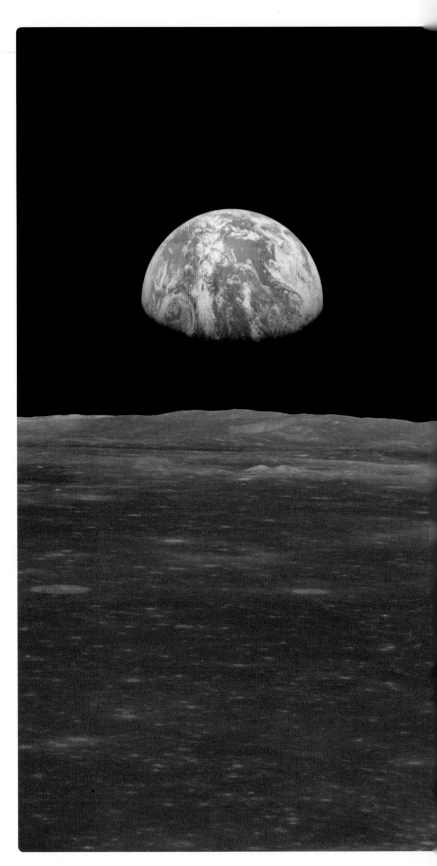

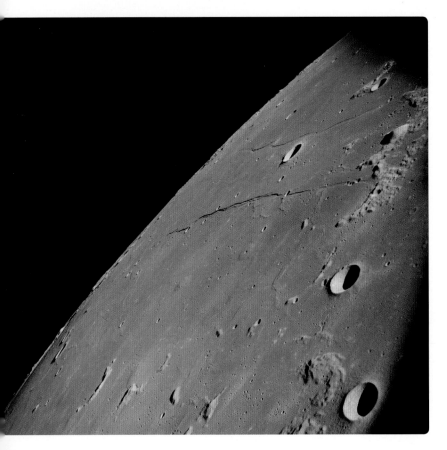

42

CRATERS GALORE

Craters pepper the lunar landscape, though there are fewer in *maria*. The biggest craters are more than 150 miles (250 km) across. These have raised walls that sink down in terraces to a smooth floor below the surrounding ground level. In the center of the floor there is typically a small mountain range. In 1998, the *Lunar Prospector* probe detected large quantities of ice in deep craters near the Moon's poles. Some of the newest craters show up well at the time of the full moon. They display bright streaks radiating like the spokes of a wheel. They are called "crater rays." The ones around Tycho are particularly prominent.

LUNAR ORIGINS

The most favored view of the Moon's origin is that it formed when a large body collided with Earth just after the planets formed some 4,500 million years ago. The Moon formed from Earth's rocky material that had been knocked into orbit.

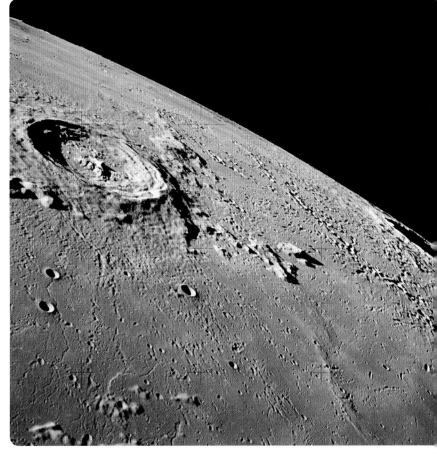

▶ The craters Eratosthenes (left) and Copernicus (right) in the region lying to the east of the Ocean of Storms.

▼ Below left: Looking down on the Sea of Tranquility.

▼ Below Right: A small crater in the Ocean of Storms.

ON THE SURFACE

The most common kinds of rocks on the Moon are *basalt* and *andesite*, both of which are seen on Earth. Dark basalt covers the *maria* while andesite is in the older highland regions. The Moon's surface also has *breccia*, which is made up of chips of old rocks stuck together. This type of rock forms when meteorites crash on the surface and smash existing rocks to bits. The surface of the Moon is covered by *regolith*, a soft dust made from powdered rocks.

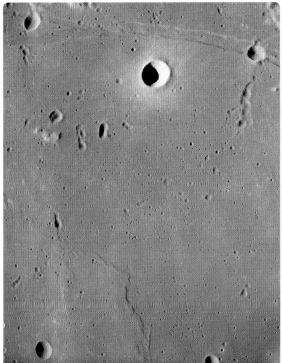

Profiling the Planets

The eight planets of the solar system are remarkably different from one another. Mercury is a barren, cratered world, baked oven-hot by the nearby Sun. Venus is even hotter, with an oppressive, deadly atmosphere. Earth is the jewel of the solar system, a colorful world of blue seas and verdant continents, home to millions of species of plants and animals in infinite variety. The red planet, Mars, is a cold wilderness with only a whiff of atmosphere.

These four planets lie relatively close together in the heart of the solar system. Then there is a huge gap of nearly 350 million miles (550 million km) before we come to the next planet, Jupiter. This is the first and largest of the four gas giants of the outer solar system. Jupiter has colorful bands of clouds in a fast-moving and stormy atmosphere. Saturn delights us with its beautiful, shining rings. Uranus and Neptune are blue-looking worlds that are alike in size and composition. Far-distant Pluto is in a class by itself, a deep-frozen ball of ice and rock.

MERCURY

Diameter at equator:	3,031 miles (4,878 km)
Avg. distance from Sun:	36 million miles (58 million km)
Spins on axis in:	58 days 16 hours
Circles Sun in:	88 days
Moons:	0

VENUS

Diameter at equator:	7,521 miles (12,104 km)
Avg. distance from Sun:	67 million miles (108 million km)
Spins on axis in:	243 days
Circles Sun in:	224.7 days
Moons:	0

EARTH

Diameter at equator:	7,926 miles (12,756 km)
Avg. distance from Sun:	93 million miles (149.6 million km)
Spins on axis in:	23 hours 56 minutes
Circles Sun in:	365 ¼ days = 1 year
Moons:	1

MARS

Diameter at equator:	4,222 miles (6,794 km)
Avg. distance from Sun:	142 million miles (228 million km)
Spins on axis in:	24 hours 37 minutes
Circles Sun in:	687 days
Moons:	2

JUPITER

Diameter at equator:	88,800 miles (143,000 km)
Avg. distance from Sun:	484 million miles (778 million km)
Spins on axis in:	9 hours 55 minutes
Circles Sun in:	11.9 years
Moons:	69

◀ The planet Jupiter with its moon Io. Named after the Roman king of the gods, Jupiter is an appropriate name for this undisputable king of the planets.

SATURN

Diameter at equator:	72,900 miles (120,000 km)
Avg. distance from Sun:	887 million miles (1,427 million km)
Spins on axis in:	10 hours 33 minutes
Circles Sun in:	29.5 years
Moons:	62

▼ Saturn, a favorite planet of astronomers, sits amid shining rings that show up in telescopes.

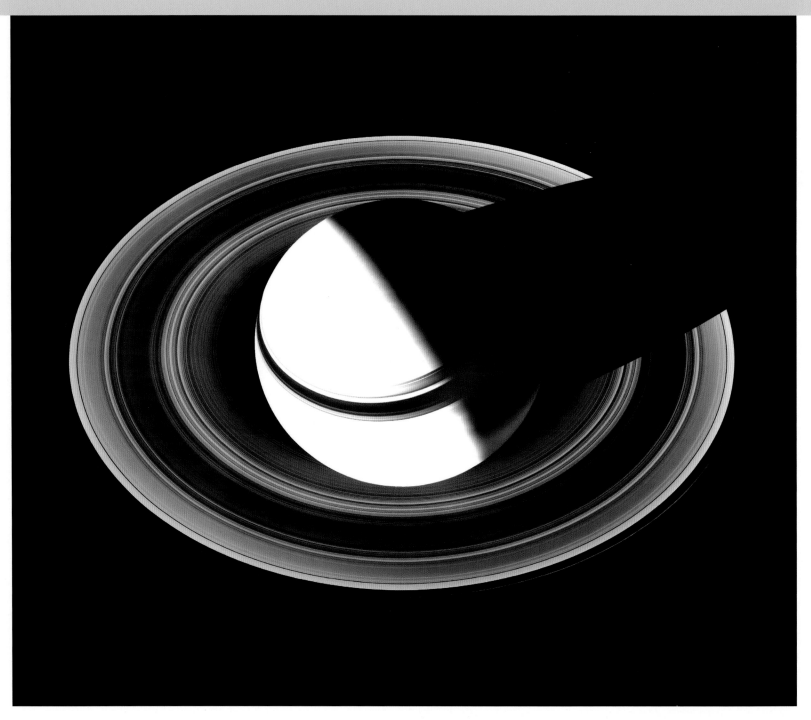

URANUS

Diameter at equator:	31,800 miles (51,200 km)
Avg. distance from Sun:	1,783 million miles (2,870 million km)
Spins on axis in:	17 hours 14 minutes
Circles Sun in:	84 years
Moons:	27

NEPTUNE

Diameter at equator:	30,800 miles (49,500 km)
Avg. distance from Sun:	2,794 million miles (4,497 million km)
Spins on axis in:	16 hours 7 minutes
Circles Sun in:	165 years
Moons:	14

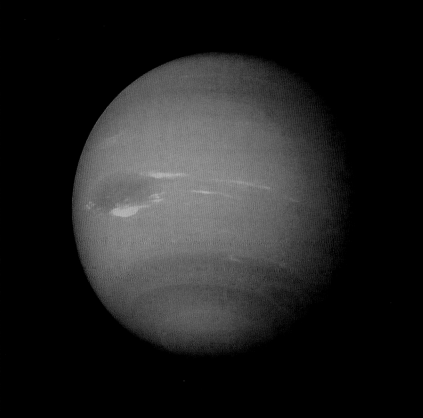

Part Two
The Star Maps

Spotting the Constellations

Although there is no actual celestial sphere, astronomers assume there is one when they come to map the heavens and locate the stars. They pinpoint a star on the sphere by a similar system to that used in geography to pinpoint a place on the surface of Earth—latitude and longitude.

In geography, the latitude of a place is a measure of the distance that place is north or south of the Equator. Its longitude is a measure of how far around the world it is from a fixed point. The latitude and longitude are both measured in degrees. Latitude is measured in degrees from the Equator, longitude from a north–south line, or "meridian," running through Greenwich in England. It is called the Greenwich Meridian.

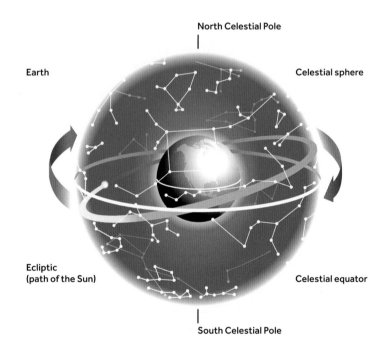

Earth

North Celestial Pole

Celestial sphere

Ecliptic
(path of the Sun)

Celestial equator

South Celestial Pole

CELESTIAL LONGITUDE

Celestial longitude, like its terrestrial counterpart, is a measure of the distance around the celestial sphere a star is from a fixed point. In this case the fixed point is the place where the Sun appears to travel across the celestial equator in the spring. Or, in other words, it is the point where the ecliptic (path of the Sun) and the celestial equator intersect. This sounds more complicated than it is. The diagram will help make it clearer.

The point of intersection is known as the First Point of Aries (A). The celestial longitude of a star is measured from this point around the celestial equator to a great circle passing through the star and the celestial poles. Again, this sounds complicated, but the diagram will clarify it.

Celestial longitude is referred to as "right ascension" (RA). It is not measured in degrees, but in hours, and these are not the everyday hours that we set our clocks by, but hours of astronomical or sidereal time—time relative to the stars! (The subject of normal and sidereal time is also discussed on page 31.)

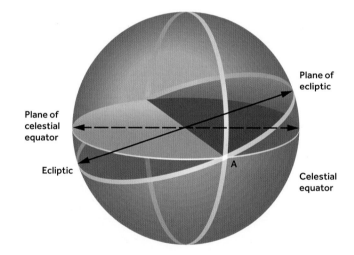

Plane of ecliptic

Plane of celestial equator

Ecliptic

A

Celestial equator

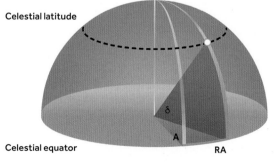

Celestial latitude

Celestial equator

First Point of Aries

δ

A

RA

CELESTIAL LATITUDE

The system of celestial latitude and longitude works on the same principles. Celestial latitude is a measure of how far a star is north or south of the celestial equator. The celestial equator is a circle around the middle of the celestial sphere, just as the Equator is a circle around the middle, or "waist," of Earth. Celestial latitude is called declination (symbol δ, Greek delta). It, too, is measured in degrees. Degrees north of the equator are deemed positive, while those south of the equator are negative.

The Star Maps

The stars move overhead during the night because Earth spins around in space. They rise over the horizon in the east, culminate (climb highest) at the meridian, and eventually set beneath the horizon in the west. As the months go by, different constellations appear in these whirling skies, while others disappear. This happens because of Earth's other motion in space—its orbit around the Sun.

Every month, as Earth travels a bit farther in its orbit, we look out at night onto a slightly different part of the celestial sphere. After a year, Earth comes full circle as it completes its orbit around the Sun, and we look out at the same sky that we saw 12 months before.

In this book we follow the changing aspects of the night sky month by month, showing how the skies look at the same time of night.

THE SKETCH MAPS

For each monthly section, we present sets of sketch maps to help stargazers find their celestial bearings. They feature the most prominent constellations.

There are four sketch maps—two for observers in the Northern Hemisphere and two for observers in the Southern Hemisphere. One of each pair shows the skies looking north; the other shows the skies looking south. If connected together, they would present a 360-degree view of the heavens that month.

The coverage of the stars on the celestial sphere is completed by two Polar maps, one in the north and one in the south. Many observers viewing the skies in the Northern Hemisphere can see some of the North Polar stars all the time. We say these stars are "circumpolar." Likewise, some of the South Polar stars are circumpolar for many Southern Hemisphere observers.

January Stars

TOP: LOOKING SOUTH

Constellations visible in North America and Europe at about 11pm on or near January 7

Orion appears in mid-skies nearly due south, and as ever, the mighty hunter serves as an incomparable signpost to a host of first-magnitude stars. Southeast is the brightest star in the sky: Sirius, the Dog Star. Northwest is Aldebaran, the red eye of the charging bull, and beyond it, the magnificent Pleiades cluster. Northeast are the twins Castor and Pollux. Due east is Procyon.

BOTTOM: LOOKING NORTH

Constellations visible in North America and Europe at about 11pm on or near January 7

As in all the northern hemisphere views looking north, Polaris, or the North Star, sits directly on the meridian. Circling counterclockwise around it are the circumpolar constellations that stay visible all the time. In this January view from mid-latitudes, Cassiopeia, Cepheus, Draco, and Ursa Minor and Major (the Little and Big Dippers, or Plough) are the circumpolar constellations.

TOP: LOOKING SOUTH

Constellations visible in Australia and South Africa at about 11pm on or near January 7

Unlike the Northern Hemisphere, the Southern Hemisphere has no convenient pole star, but the long axis of Crux, the Southern Cross, acts as a passable pointer to the Southern Celestial Pole. Around this pole, the southern stars rotate in a clockwise direction.

BOTTOM: LOOKING NORTH

Constellations visible in Australia and South Africa at about 11pm on or near January 7

The unmistakable figure of Orion appears high in the sky nearly due north. As ever, the mighty hunter serves as an incomparable signpost to a host of first-magnitude stars. Due east of Betelgeuse is Procyon. Northeast are the twins Castor and Pollux. Low down near the northern horizon is brilliant Capella. Northwest is Aldebaran, the red eye of the charging bull, and beyond it, the magnificent Pleiades cluster.

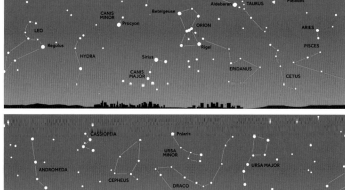

NORTHERN HEMISPHERE

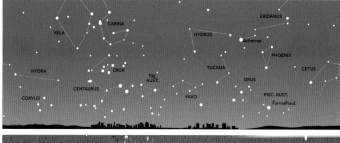

SOUTHERN HEMISPHERE

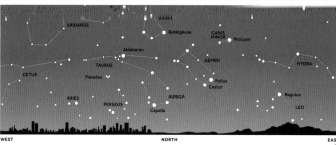

THE MONTHLY MAPS

The main monthly star maps are drawn using a grid of celestial latitude and longitude, that is, degrees of declination on the vertical axis and hours of right ascension on the horizontal axis. The maps show the constellations that appear near the meridian—the north–south line in the sky—at about 11pm in the first week of the month. They may appear slightly to the east or west, depending on the exact date of observation.

The stars shown in each map are ones that can be seen with the naked eye. The various sizes of dots representing stars give an indication of their brightness. The largest represent the brightest, first-magnitude stars. The smallest represent stars of about the fourth magnitude.

Stars may be identified in a number of ways. Some of the most prominent have proper names, such as Vega, which is the brightest star in the constellation of Lyra. Others are usually identified by a letter of the Greek alphabet, the brightest in a constellation being Alpha (α), the next brightest Beta (β), and so on.

Certain symbols are used to identify what are called "deep-sky" objects, which include nebulae, open and globular clusters, and galaxies. Such objects are identified in two main ways, either by an M number or an NGC number. M stands for Messier and the number refers to the object's position in a famous catalog drawn up by the French astronomer Charles Messier. NGC stands for *New General Catalogue*, a listing of nebulae and clusters compiled by the Danish astronomer John Louis Emil Dreyer and published in 1888.

◀ **A wide view of the constellation Cygnus, in the Milky Way.**

▶ **Map of July skies showing typical features.**

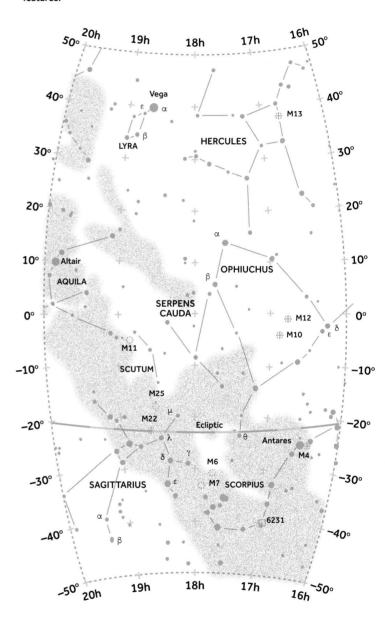

THE GREEK ALPHABET

α	Alpha	ι	Iota	ρ	Rho
β	Beta	κ	Kappa	σ	Sigma
γ	Gamma	λ	Lambda	τ	Tau
δ	Delta	μ	Mu	υ	Upsilon
ε	Epsilon	ν	Nu	φ	Phi
ζ	Zeta	ξ	Xi	χ	Chi
η	Eta	ο	Omicron	ψ	Psi
θ	Theta	π	Pi	ω	Omega

MAP SYMBOLS KEY

⬤	Star
·	Size indicates level of brightness
⬡	Open cluster
✳	Globular cluster
▢	Nebula
⬭	Galaxy
▒	Milky Way

North Polar Stars

Far northern skies lack the brilliance of far southern skies, but they do possess something that the southern ones lack—a convenient Pole Star. Astronomers call this star Polaris; another name for it is the North Star. This star is located nearly directly over Earth's North Pole, in line with Earth's axis of spin, and for this reason it hardly changes its position at all.

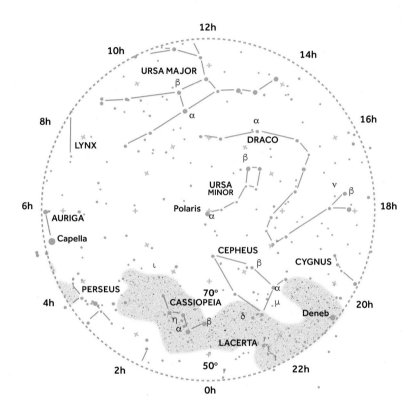

CASSIOPEIA

Cassiopeia is easy to recognize because of its distinctive "W" shape. But more than a little imagination is needed to picture this constellation, as the ancient Greeks did, as a "lady in a chair." The lady in question was Queen Cassiopeia, wife of King Cepheus, whose vanity led to their daughter Andromeda being offered as a sacrifice to the sea god Poseidon.

Cassiopeia has a number of double stars, including Alpha (α) and Eta (η), while Iota (ι) is a triple star, visible in small telescopes. But, being in the Milky Way, the constellation is most notable for its dazzling array of clusters. Some are easily picked up in binoculars, such as NGC457 and NGC654, which are found south of the W.

CEPHEUS

This is found next to the brighter Cassiopeia. In mythology, too, Cepheus and Cassiopeia were close, being king and queen of ancient Ethiopia. Their daughter was Andromeda.

Among the stars, Beta (β) shows up as a double star in small telescopes. Delta (δ) is also double; the main star is a yellow supergiant, which varies in brightness as regular as clockwork, with magnitudes between 3.5 and 4.4 in precisely five days, nine hours. It is the

prototype for the variable stars known as the Cepheids. An interesting thing about a Cepheid is that its period of variation is directly related to its absolute magnitude, or true brightness. So when we see one, we can work out how far away it is.

Mu (μ) also changes in brightness. It varies between the third and fifth magnitudes over a period of about two years; it is one of the Mira-type variables. Mu has a lovely red hue, earning it the name of Garnet Star.

DRACO, THE DRAGON

Draco is a sprawling constellation of faint stars that winds nearly halfway around the Pole Star. It is named after the dragon in Greek mythology that guarded the golden apples in the garden of the Hesperides, which Hercules collected as one of his labors.

Of interest among the stars is Nu (ν) in the Dragon's mouth, a wide double for binoculars and small telescopes. Alpha (α), also called Thuban, is notable because in ancient Egyptian times it was the Pole Star.

▲ The blue supergiant star Kappa Cassiopeiae is classed as a runaway star due to it moving through space at an unusually high velocity relative to the surrounding interstellar medium.

▶ The star-forming cloud Cepheus B.

URSA MAJOR, THE GREAT BEAR

This group features as a Key Constellation (see page 58).

URSA MINOR, THE LITTLE BEAR

Ursa Minor is also known as the Little Dipper because it looks rather like a ladle. In mythology, the Little Bear represents a nymph who nursed Zeus as a baby. It is a small constellation, which contains Polaris.

Ursa Major: The Great Bear

This is the third largest constellation. It includes the prominent asterism—group of stars—known as the Big Dipper in North America, because it resembles a ladle used to dip into a bucket. That distinguishes it from the similarly shaped Little Dipper, the constellation Ursa Minor. In Europe this star group is seen as a plough, taking the form of the handle and share blade of an old-fashioned horse-drawn plough.

To the Greeks, Ursa Major was Callisto, a beautiful girl who was seduced by the king of the gods, Zeus. After Callisto bore Zeus's child, his wife Hera turned her into a bear.

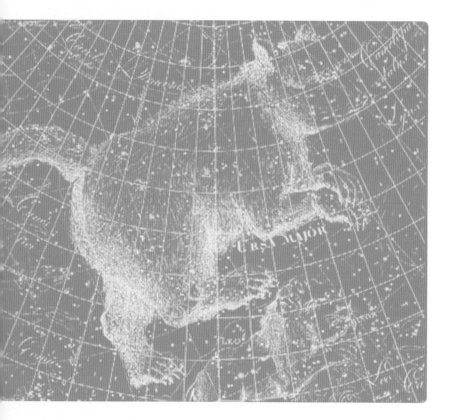

POINTING NORTH

The Big Dipper, or Plough, has been a boon to navigators for thousands of years. This is because two of its stars point to the North Star, or Polaris. The two Pointers are named Merak and Dubhe.

A line from Merak through Dubhe will point to Polaris and always show the direction north. The best-known other star in the Big Dipper (Plough) is the second along the handle, named Mizar. Close by is another bright star, Alcor. The two stars form one of the best-known double stars in the heavens. They make for a good eye test—if you can see them separately, your eyesight is good.

Ursa Major is located in a relatively featureless part of the heavens, which is why it stands out so prominently. But it has a few interesting objects, including a few galaxies within range of binoculars. They include M101, a spiral found by following the line of three stars in the Big Dipper or Plough handle. Northeast of Dubhe is the beautiful and bright spiral M81, a galaxy similar to our own. Seemingly entangled with it is M82.

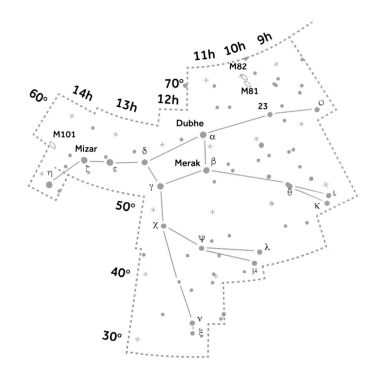

▶ Facing page: This spiral galaxy is M106, one of many to be seen near the constellation.

South Polar Stars

In the far south there is no convenient pole star as there is in the far north, so it is a little more difficult for people to get their bearings. However, the south polar region is dazzlingly bright, containing the most brilliant region of the Milky Way and beacon stars like Alpha and Beta Centauri, Canopus, and Achernar.

CARINA, THE KEEL

This constellation was once part of the ancient constellation of Argo Navis, the ship of the Argonauts. It lies on the edge of the Milky Way and is rich in nebulae and clusters. Its leading star, Canopus, is the second brightest star in the heavens, after Sirius.

Among the dazzling array of stars in the constellation, Eta (η) is special. At present it is just visible to the naked eye, but only a century ago it blazed brighter than any other star in the sky except for Sirius. It is embedded in a glorious nebula, which is visible to the naked eye but best seen in binoculars or small telescopes.

The stars Iota (ι) and Epsilon (ε) lie close to Delta (δ) and Kappa (κ) in Vela. This quadrangle of stars forms an X-shape known as the False Cross, because it may be mistaken for the Southern Cross, Crux.

Among the many bright clusters that Carina boasts, IC2602 is the finest, located around Theta (θ). It has eight stars brighter than the sixth magnitude, and is well called the Southern Pleiades.

CENTAURUS, THE CENTAUR

This dazzling group of stars features as a Key Constellation (see page 62).

CRUX, THE SOUTHERN CROSS

The Southern Cross is the most famous southern constellation of all, even though it is the smallest in the sky. The ancients regarded it as part of the hind legs of Centaurus, and it has been recognized as a separate constellation only since the 1600s.

In the Cross itself, Gamma (γ) is reddish, contrasting with the three other stars, which are bluish white. It is a double, like Alpha (α), also named Acrux, and Beta (β). Around Kappa (κ), which is close to Beta, is one of the finest open clusters in the heavens, full of sparkling colors. Designated NGC4755, it is better known as the Jewel Box.

Between Kappa and Alpha there is a dark void in the starry vista, known as the Coal Sack. It is not a hole in the Milky Way, but a cloud of dark gas that blots the light from the stars behind.

DORADO, THE SWORDFISH

Another relatively modern (1600s) constellation, Dorado is best known for hosting the nearest galaxy to our own, the Large Magellanic Cloud (LMC). This irregular galaxy can be seen with the naked eye as a fuzzy patch at the southern end of Dorado's line of stars. The LMC's brightest nebula, the Tarantula, is also visible to the naked eye.

TUCANA, THE TOUCAN

This southern bird, also introduced in the 1600s, plays host to a companion of the Large Magellanic Cloud, the Small Magellanic Cloud (SMC). This too is an irregular galaxy. The Toucan's other claim to fame is the naked-eye object originally classed as star 47—telescopes reveal this to be a gigantic ball containing hundreds of thousands of stars.

▲ NGC 299, an open star cluster located in Tucana.

▶ Top: A little-known spiral galaxy in Dorado.

▶ Bottom: Crux, the Southern Cross.

Centaurus: The Centaur

Centaurus is a spectacular feature of far southern skies. It surrounds Crux, the Southern Cross, and is named after one of the wise centaurs in Greek mythology, named Chiron. Centaurs had the torsos and heads of men, but the backs and legs of horses.

THE SOUTHERN POINTERS

The two brightest stars in Centaurus are unmistakable pointers to Crux. The brighter of the two, Alpha (α) Centauri, is the third most brilliant star in the heavens, after Sirius and Canopus. It is also the nearest bright star to us, being some 4.3 light-years away. Small telescopes will show that Alpha Centauri is a fine binary star, made up of two components. Close by is a much fainter third star, known as Proxima Centauri. This red dwarf is the closest star to us, at a distance of about 4.2 light-years.

A fainter star in the constellation, Omega (ϖ) Centauri, is not a star at all, but a globular cluster, a congregation of hundreds of thousands of stars, just one of hundreds that orbit around the center of our galaxy. Omega Centauri can readily be seen with the naked eye, and binoculars will begin to show what a huge object it is. Telescopes will reveal how packed together the stars are in the cluster. In the center, they are probably only about a tenth of a light-year apart.

Close to the triplet of stars in the left shoulder of the figure of the Centaur is an object easily visible in binoculars. It is a galaxy (NGC5128), which appears to be cut through by a dark dust lane. Radio astronomers have discovered that it is a powerful source of radio waves, called Centaurus A. Indeed, there are only two more powerful sources in the sky. Centaurus A emits so much energy that it is classed as an active galaxy, and is probably powered, like other active galaxies, by a black hole.

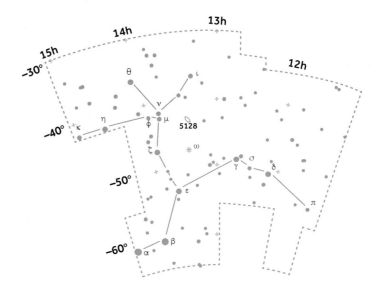

▶ Facing page: Some of the stars and billowing gas clouds near the center of the Centaurus A radio galaxy, pictured by the Hubble Space Telescope.

January Stars

January skies are probably the most stunning of the year, dominated by the majestic figure of Orion. We look at this fine constellation in detail on page 66. But its two brilliant stars, Betelgeuse and Rigel, are only two of the month's highlights. In north and south alike, brilliant stars shine down—Arcturus, Castor, Pollux, Aldebaran, Procyon, and dazzling Sirius. The Milky Way runs right across the sky, providing a feast for viewers using binoculars.

AURIGA, THE CHARIOTEER

This bright northerly constellation has the easily recognizable shape of a kite. Its brightest star is the yellowish Capella, the sixth most brilliant in the heavens. Capella represents a she-goat draped over the left shoulder of the charioteer. Just to the south are three of the goat's kids, visible as a triangle of fainter stars.

Of the other stars, Epsilon (ε) is most interesting. It is an eclipsing binary that consists of a brilliant supergiant star and a dark companion revolving around each other. Every 27 years the dark companion passes in front of the supergiant, which has the effect of making Epsilon dim noticeably.

The Milky Way runs through the constellation and features a number of bright star clusters that are visible with binoculars.

CANIS MAJOR, THE GREAT DOG

This constellation may be small, but it is easy to pick out because it boasts the brightest star in the sky. This star is Sirius, also known as the Dog Star. As stars go, Sirius is not particularly luminous. It appears very bright to us because it is close, at a distance of only about nine light-years.

One star in the constellation you won't be able to spot is the so-called Companion of Sirius, also called the Pup. Sirius and the Pup circle around each other, forming a two-star, binary system. The Pup is a white dwarf star—tiny and very dense, it was the first white dwarf to be discovered, in 1862.

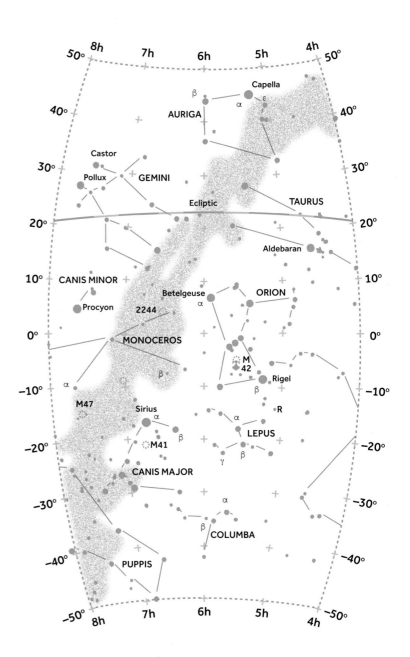

▲ Constellations visible near the meridian at about 11pm during the first week in January.

COLUMBA, THE DOVE

One of the many bird constellations found in the far Southern Hemisphere, Columba represents the dove Noah released from the Ark to search for dry land. It is relatively faint but quite easy to spot because it appears in an otherwise featureless region.

LEPUS, THE HARE

The Hare, which lies beneath the feet of Orion, is running away from the larger of his dogs, Canis Major. The star Gamma (γ) is a double, which can be seen through binoculars. One of the most interesting stars is R. It is a variable, which at its brightest can easily be seen through binoculars. It has a striking red color, and is often called the Crimson Star.

ORION

Straddling the celestial equator, this magnificent star group is this month's Key Constellation (see page 66).

Orion

The Greeks named this magnificent constellation after a mighty hunter. He was the son of the sea god Poseidon, and boasted that he could kill any creature on Earth. He strides across the heavens with a club raised in his right hand and a shield in the other. A sword dangles from his belt. At his feet are his two dogs, Canis Major and Minor.

Orion fell in love with seven sisters, the Pleiades, and still pursues them in the sky, following their star cluster as the heavens rotate. Orion himself is chased by Scorpius, the Scorpion. In myth the gods sent a scorpion to kill Orion.

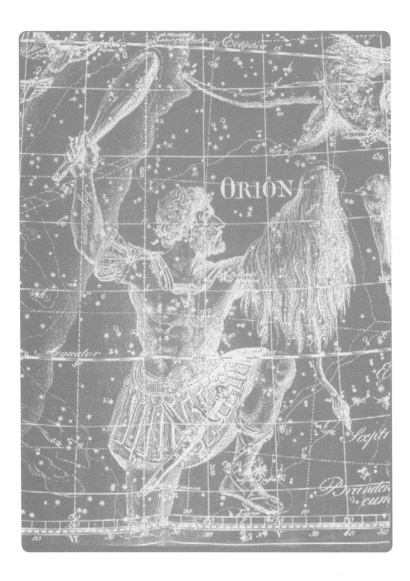

BIG AND BRIGHT

Orion is a familiar sight to stargazers of both hemispheres because it straddles the celestial equator. Its main outline is marked by seven bright stars. At the figure's left foot is Rigel, which is opposite Betelgeuse at Orion's right shoulder. The two make a contrasting pair: Rigel is brilliant white and slightly brighter than reddish Betelgeuse. Betelgeuse is a supergiant and one of the biggest stars we know, with an estimated diameter of at least 250 million miles (400 million km). If it were located where the Sun is in our solar system, it would reach out beyond Mars.

Another feature of the constellation figure is the group of three brilliant stars that form Orion's belt. To their south is the glowing patch that forms part of Orion's sword. It is a vast region of glowing gas, known as the Great Nebula, the Orion Nebula, or M42. It is lit up by four stars in a multiple-star system called the Trapezium, embedded within it. The Nebula is a hotbed of star formation, as are other nebulous regions in the constellations. Another located close to the most southerly star in Orion's belt (Zeta ζ) bears an uncanny resemblance to the head and neck of a horse, which is why it is named the Horsehead Nebula.

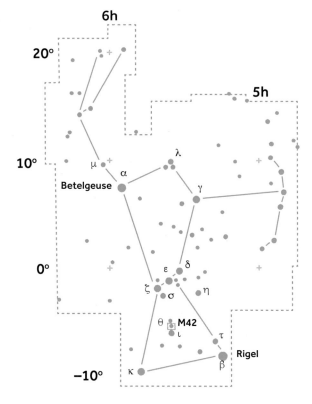

▶ The glorious Orion Nebula.

◀ NASA released this stunning infrared image of the Horsehead Nebula to celebrate Hubble's 23rd year.

▶ NGC 2174, also known as the Monkey Head Nebula, lies in the constellation of Orion—a colorful region of young stars, cosmic gas, and dust.

January Stars

TOP: LOOKING SOUTH

Constellations visible in North America and Europe at about 11pm on or near January 7

Orion appears in mid-skies nearly due south, and as ever, the mighty hunter serves as an incomparable signpost to a host of first-magnitude stars. Southeast is the brightest star in the sky: Sirius, the Dog Star. Northwest is Aldebaran, the red eye of the charging bull, and beyond it, the magnificent Pleiades cluster. Northeast are the twins Castor and Pollux. Due east is Procyon.

BOTTOM: LOOKING NORTH

Constellations visible in North America and Europe at about 11pm on or near January 7

As in all the northern hemisphere views looking north, Polaris, or the North Star, sits directly on the meridian. Circling counterclockwise around it are the circumpolar constellations that stay visible all the time. In this January view from mid-latitudes, Cassiopeia, Cepheus, Draco, and Ursa Minor and Major (the Little and Big Dippers, or Plough) are the circumpolar constellations.

NORTHERN HEMISPHERE

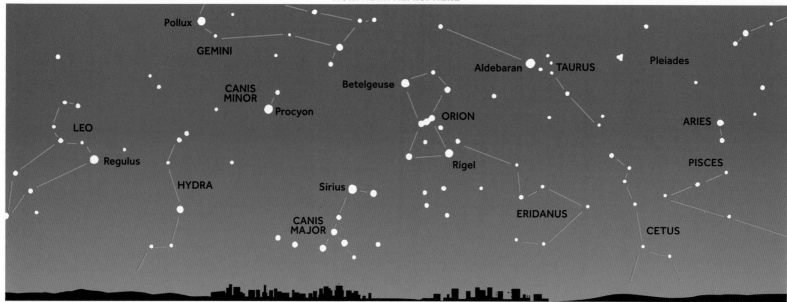

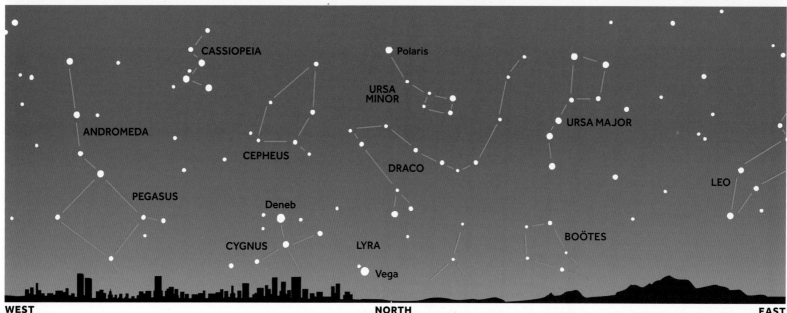

TOP: LOOKING SOUTH

Constellations visible in Australia and South Africa at about 11pm on or near January 7

Unlike the Northern Hemisphere, the Southern Hemisphere has no convenient pole star, but the long axis of Crux, the Southern Cross, acts as a passable pointer to the Southern Celestial Pole. Around this pole, the southern stars rotate in a clockwise direction.

BOTTOM: LOOKING NORTH

Constellations visible in Australia and South Africa at about 11pm on or near January 7

The unmistakable figure of Orion appears high in the sky nearly due north. As ever, the mighty hunter serves as an incomparable signpost to a host of first-magnitude stars. Due east of Betelgeuse is Procyon. Northeast are the twins Castor and Pollux. Low down near the northern horizon is brilliant Capella. Northwest is Aldebaran, the red eye of the charging bull, and beyond it, the magnificent Pleiades cluster.

SOUTHERN HEMISPHERE

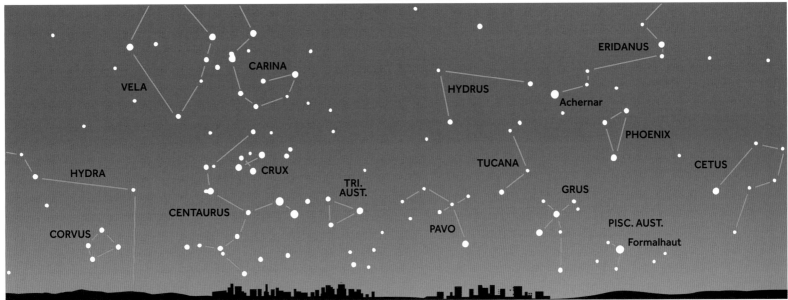

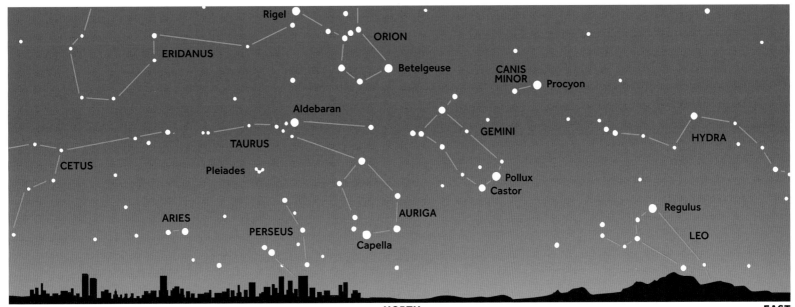

WEST

NORTH

EAST

February Stars

February skies are not as impressive as those of the previous month, since rather faint constellations are moving in from the east. Only in far southern skies do the Milky Way and bright constellations still dazzle.

CANCER, THE CRAB

This is one of the constellations of the Zodiac, located between Gemini and Leo. The Sun passes through Cancer between July 20 and August 10. Its stars are quite faint—in fact, it is the faintest constellation of the Zodiac. Nevertheless, it has several interesting features.

The most impressive is the Beehive, a fine open cluster visible to the naked eye but better seen with binoculars. It was given this name because astronomers likened the clusters of stars to bees busily buzzing around their hive. Its proper astronomical name is Praesepe, or M44.

CANIS MINOR, THE LITTLE DOG

This tiny constellation represents the smaller of the two dogs that accompany Orion. The constellation has only two bright stars, the brighter being named Procyon, which means "before the dog."

This refers to the fact that it rises before the Dog Star, Sirius. Procyon is the eighth-brightest star in the heavens, and one of the nearest, at a distance of about 11 light-years.

GEMINI, THE TWINS

This fine constellation is found in the Zodiac between Cancer and Taurus. The Sun passes through it between June 21 and July 20. Its pair of bright main stars, both of the first magnitude, are named Castor and Pollux. To the ancient Greeks, they represented the twin sons, hatched from an egg laid by Leda, Queen of Sparta, after she had been wooed by Zeus (see page 102). To the ancient Romans, they represented the twins Romulus and Remus, the legendary founders of Rome, who were raised by a wolf.

Of the twins, Castor is slightly fainter, but is also the more interesting, being an impressive multiple-star system. Small telescopes will reveal that it is a binary system, consisting of two blue-white stars circling around each other.

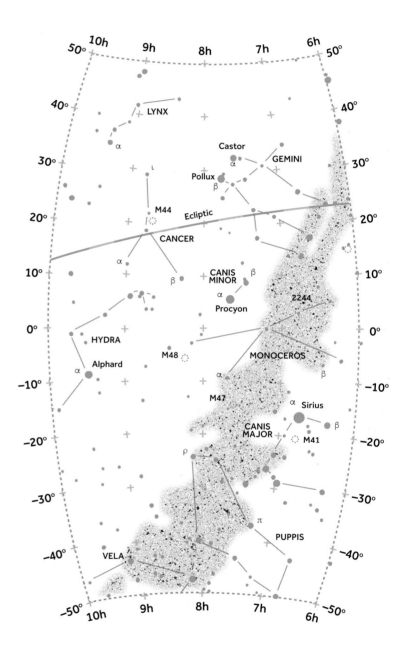

▲ Constellations visible near the meridian at about 11pm during the first week in February.

▲ Left: One of the gems in Monoceros, the lovely Rosette Nebula, with petal-like skeins of gas.

▲ Right: ESO 255-7, a quartet of interacting galaxies in the constellation Puppis.

MONOCEROS, THE UNICORN

This constellation has only relatively faint stars, but straddles a particularly rich region of the Milky Way. Beta (β) is a multiple-star system, with three components, easily seen as a triangle in a small telescope. The crowning glory is the magnificent Rosette Nebula. It is a circular cloud of red gas, with petal-like regions surrounding a cluster of brilliant stars (NGC2244).

PUPPIS, THE POOP DECK OR THE STERN

Puppis and the neighboring Vela and Carina constellations were once included in a much larger constellation called Argo Navis. It represented the ship in which the Argonauts sailed to search for the Golden Fleece. Puppis represents the stern of the ship, Vela the sails, and Carina the keel.

Much of this brilliant southern constellation lies within the Milky Way, so it is rich in star clusters and nebulae. Brightest of the clusters is M47, easily visible to the naked eye in the Milky Way to the east of Sirius.

February Stars

TOP: LOOKING SOUTH

Constellations visible in North America and Europe at about 11pm on or near February 7

Mid-skies have lost some of their spectacle, with Orion and Taurus slipping away west, and with them the pearly white band of the Milky Way. Canis Minor and its brightest star Procyon are now close to the meridian. Castor and Pollux are high above them, while Sirius shines brilliantly closer to the horizon. In lower northern latitudes, observers may be able to glimpse the far southern constellations Puppis and Vela on the horizon.

BOTTOM: LOOKING NORTH

Constellations visible in North America and Europe at about 11pm on or near February 7

Cassiopeia and Cepheus are sinking lower, while the Big Dipper, or Plough, on the opposite side of Polaris, is climbing higher. On the meridian, low down on the horizon, is bright Deneb, marking the tail of the Swan (Cygnus). Just to the east is the somewhat brighter Vega. Arcturus becomes visible as Boötes climbs higher. West of the meridian, Andromeda is still visible below Cassiopeia, seen as a misty patch to keen eyes.

NORTHERN HEMISPHERE

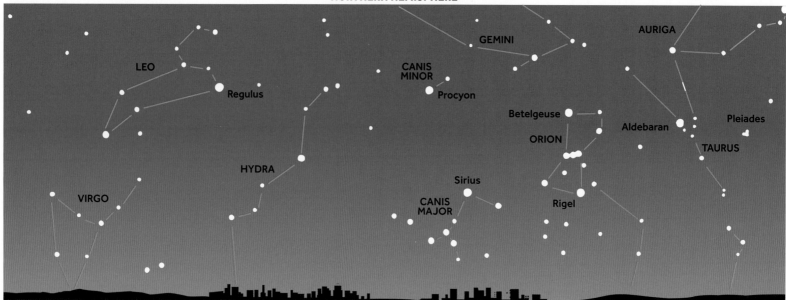

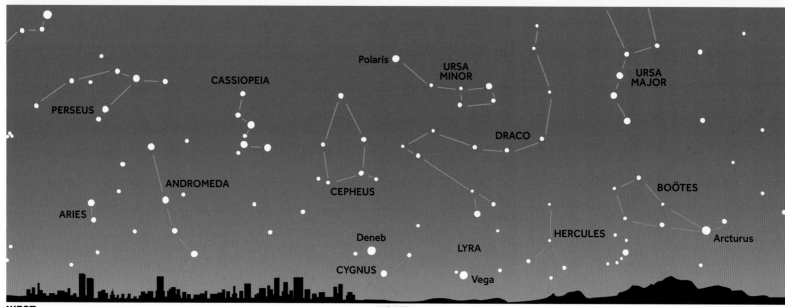

WEST NORTH EAST

TOP: LOOKING SOUTH

Constellations visible in Australia and South Africa at about 11pm on or near February 7

In February, the brightest part of the southern sky lies in the southeast, where the Southern Cross is located, its long axis nearly horizontal. As ever, it is encircled by the bright stars of Centaurus. In the far east, Spica, the lead star of Virgo, has risen. In the western half of the sky, only Achernar is conspicuous. It marks the mouth of the long river constellation Eridanus.

BOTTOM: LOOKING NORTH

Constellations visible in Australia and South Africa at about 11pm on or near February 7

Mid-skies look less spectacular as Orion and Taurus move away to the west, and with them the pearly white band of the Milky Way. Canis Minor and its brightest star Procyon are now close to the meridian. Castor and Pollux are in mid-skies below them. Northwest near the horizon, Capella is still just visible, as are the Pleiades farther west still.

SOUTHERN HEMISPHERE

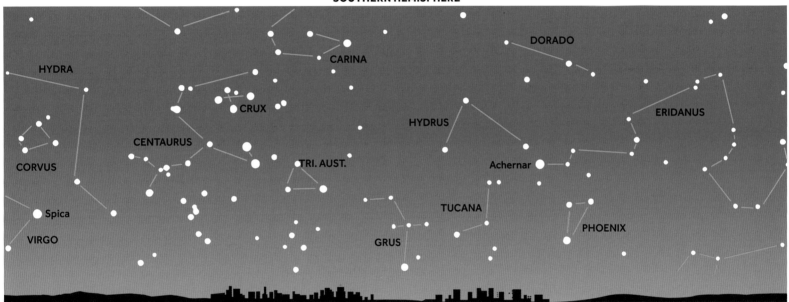

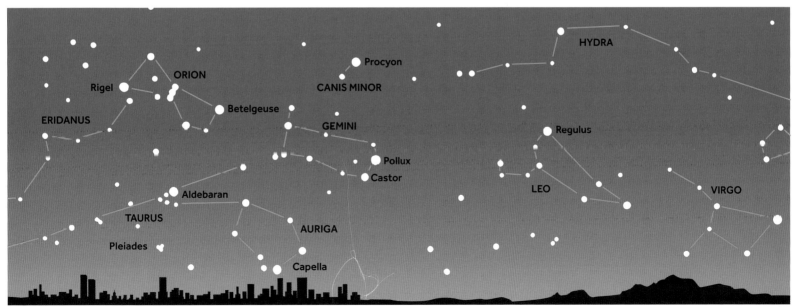

WEST   NORTH   EAST

March Stars

March skies lack the starry arch of the Milky Way. But they are boosted by the Zodiac constellation Leo, with its easy-to-recognize sickle. When Leo appears in the sky, the days are lengthening and spring is coming to the Northern Hemisphere. In the Southern Hemisphere, nights are lengthening and fall is approaching.

HYDRA, THE WATER SNAKE (HEAD)

Hydra stretches a quarter of the way around the celestial sphere and is the largest of the 88 constellations. Its undulating, serpent-like line of stars runs roughly parallel with, and to the south of, the ecliptic and the Zodiac constellations of Libra, Virgo, Leo, and Cancer. In Greek mythology, Hydra was the multi-headed snake that Hercules killed as one of his twelve labors.

Hydra may be long, but its stars are far from spectacular. It has only one star that can be considered bright, the second-magnitude orange-colored giant Alphard, a name that means the "solitary one." It is found by dropping south from the bright Regulus in Leo (See page 81 for Hydra).

LYNX, THE LYNX

This is a relatively recent constellation, first described by the Polish astronomer Hevelius in the 1600s. He chose that name as he felt that only the lynx-eyed (with sharp vision) would be able to spot it. It is faint, having only one reasonably bright star, Alpha (α), at the third magnitude.

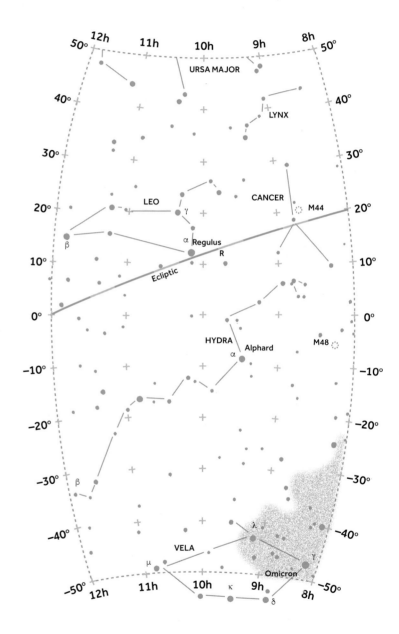

▲ Constellations visible near the meridian at about 11pm during the first week in March.

LEO, THE LION

This constellation of the Zodiac is formed by reasonably bright stars. The Sun passes through this constellation between August 10 and September 16 every year. Leo is a reasonably good likeness to the figure of a crouching lion. The curved line of stars that mark the lion's head and front is known as the Sickle, owing to its distinctive shape. In mythology, the lion was another victim of Hercules.

The lead star of the constellation is Regulus. It is also known as Cor Leonis, or Lion's Heart. It is a double star, as is another Sickle star, Gamma (γ), which is also called Algeiba, or the Lion's Mane. Still at the front end, there is the reddish Mira variable star R, which brightens to the fourth magnitude and fades to the eleventh magnitude over a period of about 10 months.

The tail end of the lion is marked by a triangle of bright stars. Beta (β), or Denebola, is a fine binary, with golden yellow components. It looks lovely in a small telescope. The tail end of the lion merges into Virgo and into a region that is rich in galaxies.

On about November 17 or 18 each year Leo plays host to a meteor shower. The radiant—the point from which the meteors appear to come—lies in the Sickle.

The source of the shower is the dust left in the wake of Comet Temple-Tuttle, which returns to the vicinity of the Sun every 33 years. As a result, showers peak over the same 33-year period; 1999 was the last good year.

VELA, THE SAILS

Once part of the large ancient Greek constellation Argo Navis, Vela is one of the splendid far-southern constellations that northern astronomers never see. It is embedded in the Milky Way and is rich in clusters and nebulae. Gamma (γ) is a multiple star that reveals up to four components in a small telescope. IC2391, around Omicron, north of Delta (δ), is an open cluster that can be seen with the naked eye, but looks much better in binoculars. Delta and Kappa (κ) are two of the stars that make up the False Cross, along with two in Carina (page 60).

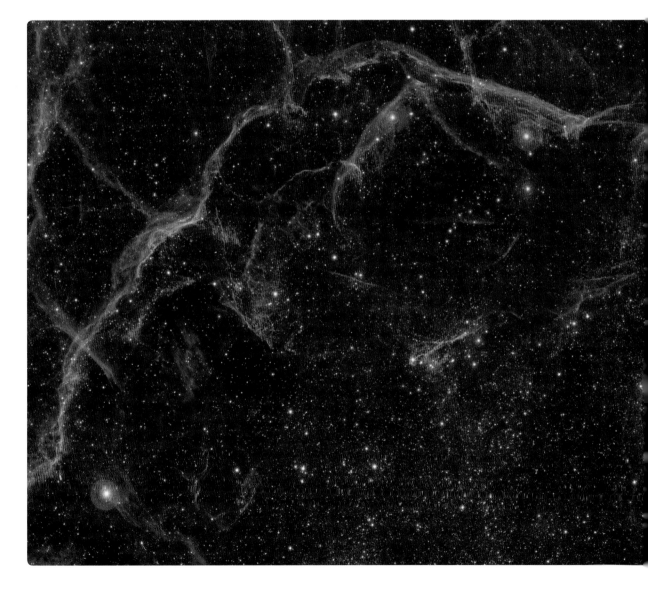

◀ A fine print of Leo, and its sidekick Leo Minor, the Little Lion. We can see that Regulus marks the big lion's heart.

▲ A long-exposure picture of the Vela Supernova Remnant, the remains of a giant star that exploded in Vela 12,000 years ago.

March Stars

TOP: LOOKING SOUTH

Constellations visible in North America and Europe at about 11pm on or near March 7

Orion and Canis Major are sinking low on the western horizon, and Rigel and Sirius will soon be setting. Mid-skies are occupied by the line of faint stars of Hydra, and higher up Leo crouches. Its brightest star, Regulus, is located due south, with the sickle-like curve of stars above it. Virgo's lead star, Spica, shines brightly in the southeast, making a prominent pair with nearby Arcturus.

BOTTOM: LOOKING NORTH

Constellations visible in North America and Europe at about 11pm on or near March 7

Cepheus is on the meridian, directly beneath Polaris. To the east and lower down are the bright pair Deneb and Vega. They are two of the three stars (the other is Altair) that will form the Summer Triangle. Continuing east, Hercules and Boötes are climbing, and sandwiched between them is the arc of stars that form the Northern Crown, Corona Borealis. On the other side of the sky, Taurus and Auriga are descending, their brilliant stars Aldebaran and Capella shining like beacons.

NORTHERN HEMISPHERE

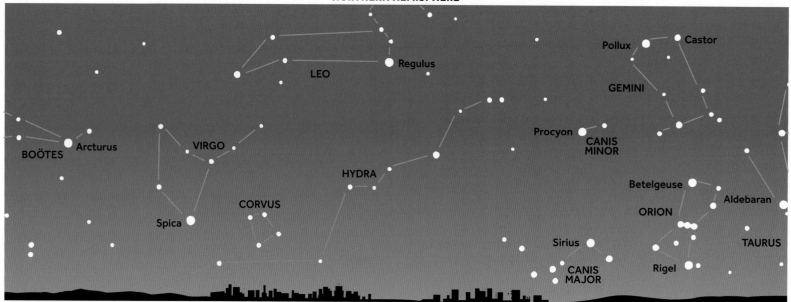

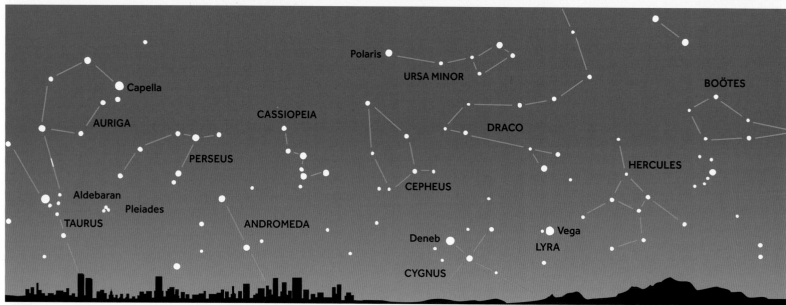

WEST   NORTH   EAST

TOP: LOOKING SOUTH

Constellations visible in Australia and South Africa at about 11pm on or near March 7

As ever, center skies are empty of really bright stars around the Celestial South Pole, but there are plenty in the southeast again. Crux and Centaurus dazzle, with Alpha and Beta Centauri now one above the other. Scorpius has risen in the east, with Antares, marking the scorpion's heart—a noticeable orange color. The Milky Way here is sumptuously rich. Across the sky in the southwest, Canopus is unmistakable in an otherwise bland region of sky away from the Milky Way.

BOTTOM: LOOKING NORTH

Constellations visible in Australia and South Africa at about 11pm on or near March 7

Orion is sinking low on the western horizon, and the brilliant Rigel and Betelgeuse will soon be setting. They form a circle of bright stars, with Castor and Pollux in nearby Gemini, and Procyon higher up. Leo is prominent in mid-skies, its brightest star Regulus located due north, with the sickle-like curve of stars below it. Virgo's lead star Spica shines brightly at about the same height in the east, making a prominent pair with Arcturus, lower down.

SOUTHERN HEMISPHERE

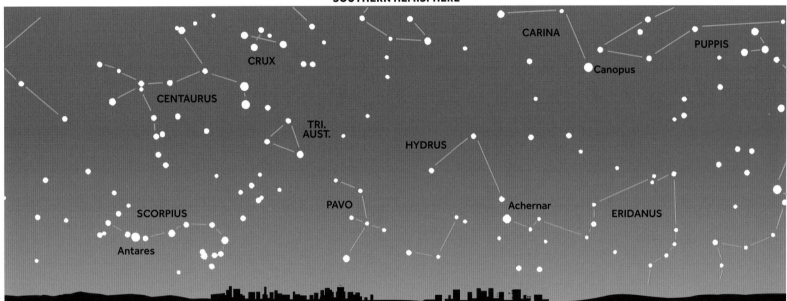

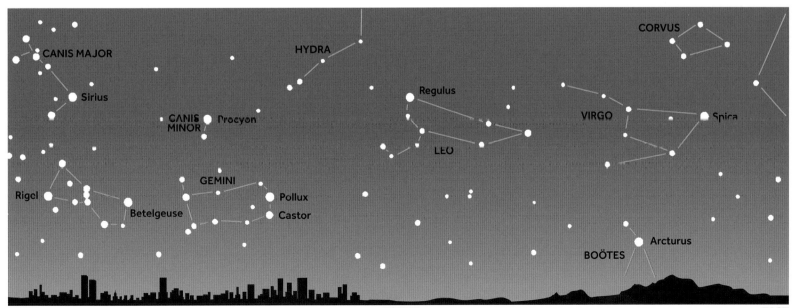

WEST   NORTH   EAST

April Stars

Leo lends some interest to the night sky, but to the south the skies are still relatively bare. This is because they are occupied by the two largest constellations—Virgo and Hydra—which possess few bright stars. Virgo at least has first-magnitude Spica, which forms a noticeable spring/fall triangle with two other equally bright stars—Regulus in Leo and Arcturus in Boötes.

CANES VENATICI, THE HUNTING DOGS

This constellation of mainly faint stars is disappointing to the naked eye, but is far more interesting in the telescope. It is a relatively modern constellation (1600s), whose subjects are supposed to be the dogs of the herdsman Boötes, used for hunting and protecting his herd from the two bears (Ursa Major and Minor).

Alpha (α) is a third-magnitude star, which in a small telescope proves to be a double. This star was given the name Cor Caroli, meaning "Charles's heart," by Edmond Halley (of the comet fame). It refers to the heart of the executed English king, Charles I.

There are two particularly interesting objects for telescopic observers. One is the Whirlpool Galaxy, M51. This is located on a line between Alpha and the first star in the handle of the Big Dipper (Plough). The Whirlpool is a particularly beautiful spiral galaxy that we see head on. The second highlight of Canes Venatici is M3, a fine globular cluster. This is found roughly halfway between Alpha and the bright Arcturus in the neighboring constellation of Boötes. It can be easily spotted with binoculars.

CORVUS, THE CROW

This small constellation is linked in mythology with Hydra (the Water Snake), and Crater (the Cup). Apollo sent the crow to fetch him water in a cup, but it stopped to eat figs and was late. In revenge Apollo sent the crow to the heavens, where, just out of reach of the cup, it is eternally thirsty.

Corvus is not a particularly interesting constellation, but is pleasing enough when viewed using binoculars. Large telescopes reveal near Gamma (γ) a pair of interacting galaxies, linked by skeins of glowing gas. They are collectively called the Antennae, because they look like the coiled antennae of a butterfly.

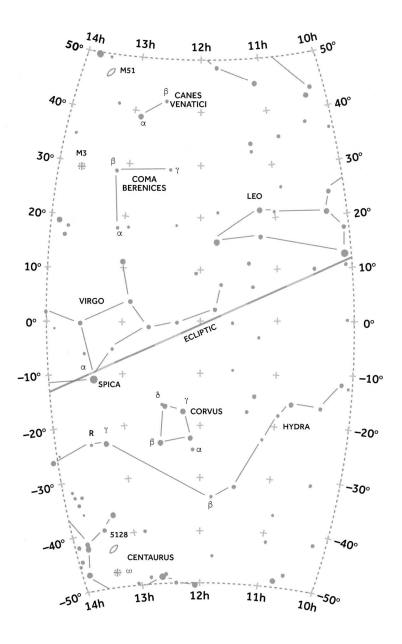

▲ Constellations visible near the meridian at about 11pm during the first week in April.

HYDRA, THE WATER SERPENT (TAIL)

Near Gamma (γ) is the reddish star R, which is a Mira-type variable. This means that it is a red giant that is easily visible with the naked eye at its brightest, but fades to a difficult binocular object at its faintest. It goes through this bright and faint cycle about every 13 months (see also page 76 for Hydra).

VIRGO, THE VIRGIN

Virgo is one of the constellations of the Zodiac, and is sandwiched between Leo and Libra. The Sun passes through Virgo between September 16 and October 31 each year. Among the constellations, Virgo covers the second-largest area of the heavens, after Hydra. However, it is not easy to make out in the sky. Only through large telescopes is Virgo really impressive.

Virgo has roots that stretch back to Babylonian times, usually in the role of a mother goddess. In Greek mythology, she was goddess of justice and sometime corn goddess. This is reflected in the name of the constellation's one outstanding star, the first-magnitude Spica, which means "ear of wheat."

▲ Top left: The incomparable Whirlpool Galaxy, M51, a head-on spiral.

▲ The Antennae galaxies in Corvus.

April Stars

TOP: LOOKING SOUTH

Constellations visible in North America and Europe at about 11pm on or near April 7

Orion has all but disappeared below the horizon in the west, where Procyon, Castor, and Pollux still shine. Leo is still prominent, just west of the meridian, and Virgo now occupies center stage. High in the east, Boötes is advancing, easily recognized by its kite shape and the brilliant Arcturus at its tail end. The rest of the sky is disappointingly bland.

BOTTOM: LOOKING NORTH

Constellations visible in North America and Europe at about 11pm on or near April 7

Cassiopeia is reaching its low point in the northern heavens, almost on the meridian. On the opposite side of Polaris (and out of the frame here), the Big Dipper, or Plough, has climbed to its highest point, with its handle roughly parallel with the horizon. In the east, Cygnus is still climbing and is now completely above the horizon, and we can fully appreciate the swanlike shape of its bright stars. The Milky Way is now almost parallel with the horizon, running from the feet of Gemini, through Cygnus.

NORTHERN HEMISPHERE

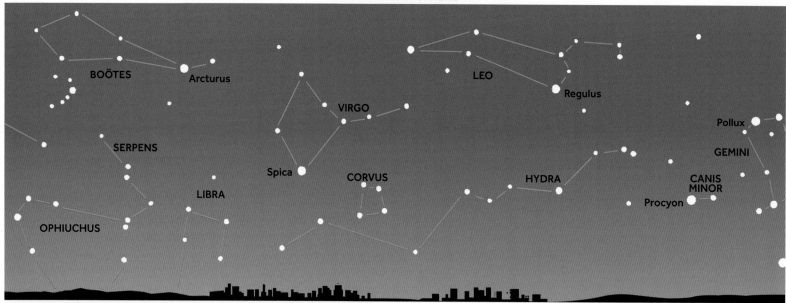

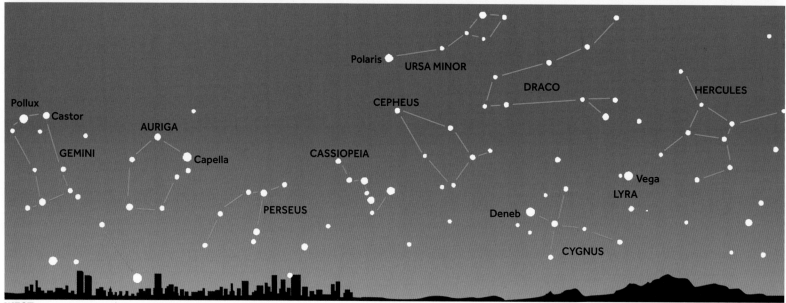

WEST   NORTH   EAST

TOP: LOOKING SOUTH

Constellations visible in Australia and South Africa at about 11pm on or near April 7

Crux has risen out of the frame here, but we know where it is by reference to the prominent pointers Alpha and Beta Centauri. It sits nearly on the meridian. Low down, the Toucan (Tucana) has reached its lowest position in the sky, while its neighbor the Peacock (Pavo) is ascending. Achernar, too, is only just above the southern horizon. Scorpius is still brightening the skies in the east, while Sirius has joined Canopus in the west.

BOTTOM: LOOKING NORTH

Constellations visible in Australia and South Africa at about 11pm on or near April 7

In the west, Orion has disappeared below the horizon, and Pollux is the only Gemini twin still visible. Procyon, too, is close to setting. Leo is still prominent, as it drifts west. Virgo is close to the meridian, but not at all easy to make out. Only Spica stands out, almost directly above Arcturus in Boötes. However, the rest of the sky is somewhat empty, with Libra, Serpens, and Ophiuchus risen in the east.

SOUTHERN HEMISPHERE

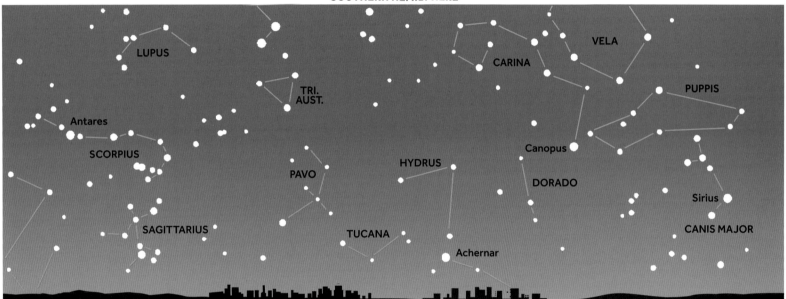

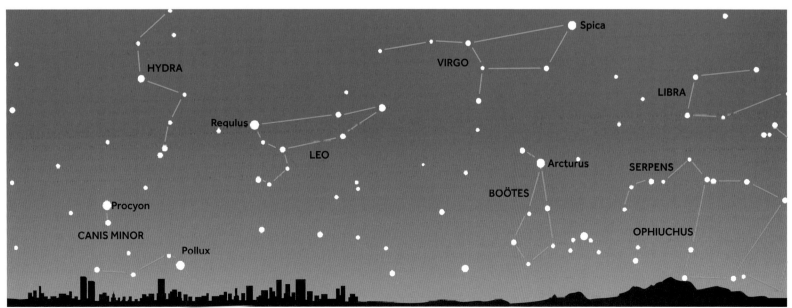

WEST — NORTH — EAST

May Stars

The two bright stars that occupy the mid-skies make an interesting contrast. The northern one is Arcturus, a noticeably orange-red giant star. The slightly dimmer southern one is Spica, also a giant but pure white. Apart from these two stars, the mid-sky region remains bland, occupied by the faint stars of Serpens (head), Libra, Virgo, and the tail of Hydra.

BOÖTES, THE HERDSMAN

This prominent northern constellation is easy to recognize by its kite shape, with bright Arcturus marking the tail end. It is located to the south of Ursa Major, the Great Bear, and is easily located by following the curve of the stars in the handle of the Big Dipper (Plough).

In Greek mythology, Boötes was Arcas, son of Callisto, who had been changed into a bear (Ursa Major). Out hunting one day, Boötes didn't recognize her and was about to kill her, but Zeus (Callisto's former lover) placed them both in the heavens to prevent this tragedy.

The bear theme is continued in the name of the constellation's lead star, Arcturus. The name means "bear guard." Arcturus is a red giant star about 20 times bigger than the Sun. Highly luminous and relatively close (25 light-years), it is the fourth-brightest star in the heavens.

Three of Boötes's bright stars are doubles, visible in a small telescope—Epsilon (ε), Xi (ξ), and Mu (μ), which binoculars will also separate. Epsilon is particularly lovely, with red and blue-green components.

CENTAURUS, THE CENTAUR

This fine far southern constellation is one of the featured Key Constellations (see page 62).

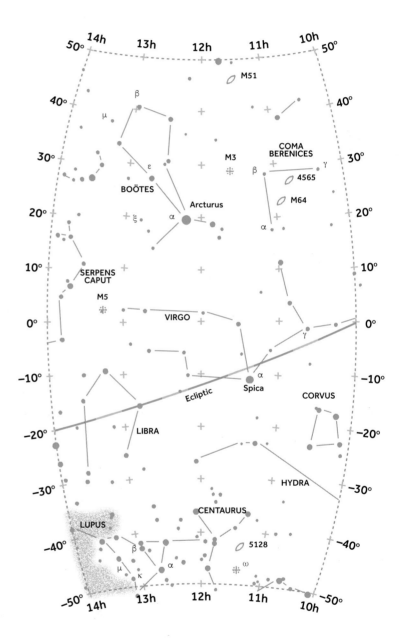

▲ Constellations visible near the meridian at about 11pm during the first week in May.

COMA BERENICES, BERENICE'S HAIR

This constellation is relatively modern (1600s) and formed out of faint stars the Greeks considered as part of Leo's tail. It represents the long tresses the Egyptian Queen Berenice cut off to thank the gods for returning her husband, King Ptolemy, safely from battle.

The only interest to the naked eye is the loose cluster of faint stars around Gamma (γ), which look much better when viewed through binoculars. Coma Berenices is a delight for telescope observers, as it is a region that is rich in galaxies. They lie to the south of Gamma and extend into the neighboring constellation of Virgo. They belong to the massive grouping of some 3,000 galaxies known as the Virgo Cluster. The majority of these galaxies typically lie more than 400 million light-years away and so are beyond the reach of most amateur telescopes. There are, however, closer galaxies within the reach of small instruments. They can be found on a line running between Gamma and Alpha (α), and include M64 and NGC4565, the Black Eye Galaxy and Needle Galaxy, respectively.

▼ Centaurus contains our nearest neighbors. The two bright stars are Alpha Centauri (left) and Beta Centauri. The faint red star in the center of the red circle is Proxima Centauri, the nearest star to the Sun.

LUPUS, THE WOLF

This constellation is located on the edge of the southern Milky Way close to Centaurus. Small telescopes reveal that many of the brightest stars in Lupus are doubles. They include Kappa (κ) and Mu (μ). Mu is an easy double, and larger instruments will show that the brighter of the pair is itself a double.

▼ The center of M100, showing clearly defined spiral arms. It is one of thousands in the Coma cluster.

May Stars

TOP: LOOKING SOUTH

Constellations visible in North America and Europe at about 11pm on or near May 7

Meridian skies are graced by Arcturus high up and Spica lower down. They contrast noticeably in color, with Spica being white, and Arcturus orange. Low down on the southeast horizon the Scorpion (Scorpius) is rising, with red Antares—marking the scorpion's heart—just visible in dark skies. In the far east another beacon star puts in an appearance, Altair in Aquila.

BOTTOM: LOOKING NORTH

Constellations visible in North America and Europe at about 11pm on or near May 7

Cassiopeia is still low in the sky, just east of the meridian. A second bird, an eagle (Aquila), has risen above the eastern horizon to join the swan (Cygnus). Their two first-magnitude stars, Deneb and Altair, form with Vega in Lyra the conspicuous Summer Triangle. Of these three, Vega appears brightest, with Deneb the least bright (as it is farther away). In absolute terms, however, Deneb is by far the most brilliant; in fact being one of the brightest stars we know in our galaxy.

NORTHERN HEMISPHERE

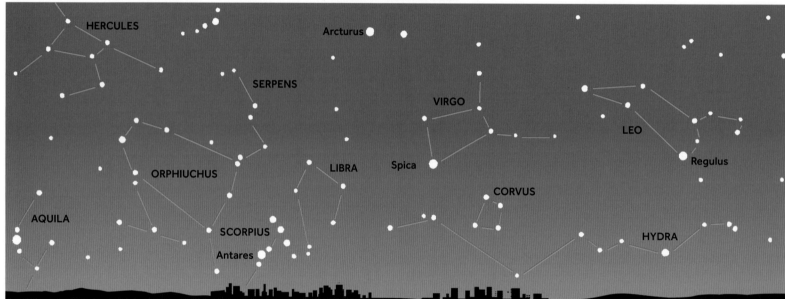

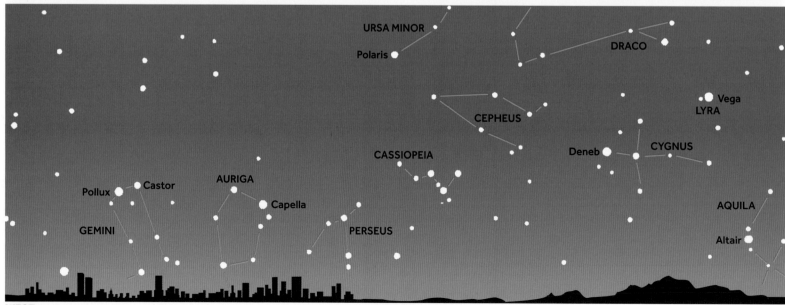

WEST   NORTH   EAST

TOP: LOOKING SOUTH

Constellations visible in Australia and South Africa at about 11pm on or near May 7

This month the Milky Way is particularly spectacular since it arches right across the sky. It stretches from Sagittarius in the east, through Canis Major, now setting in the west. Sirius has disappeared, leaving Canopus no rival, save Achernar, which is noticeably fainter. Canopus's constellation, Carina, occupies southwest skies with Vela and Puppis. In ancient times, all three of these constellations formed the single constellation Argo Navis.

BOTTOM: LOOKING NORTH

Constellations visible in Australia and South Africa at about 11pm on or near May 7

Arcturus in Boötes occupies center stage in meridian skies, with Spica in Virgo higher up (just out of the frame here). They contrast noticeably in color, with Spica being white, and Arcturus orange. High in the far east another beacon star puts in an appearance, Antares. Hercules has also risen over the horizon, but in general, the eastern sky is disappointing. The western sky is little better, with Leo soon to set.

SOUTHERN HEMISPHERE

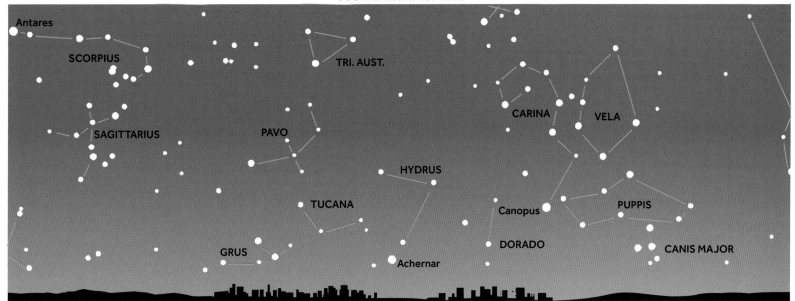

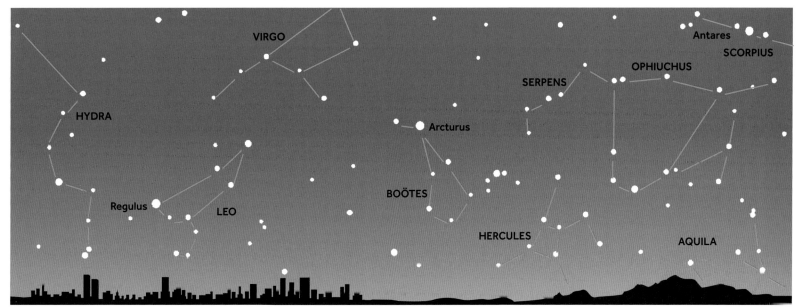

WEST   NORTH   EAST

June Stars

The skies north and south brighten overall this month, with Hercules joining Boötes in the north and Scorpius prominent in the south. Scorpius's brightest star, Antares, marking the Scorpion's heart, forms a neat triangle with Arcturus in Boötes and Spica in Virgo. It is noticeably redder than the orange Arcturus.

CORONA BOREALIS, THE NORTHERN CROWN

Corona Borealis is the crown of Princess Ariadne, daughter of King Minos of Crete. When she married Dionysus, he flung the crown into the heavens, whereupon its sparkling jewels were transformed into stars.

The constellation's brightest star, Alpha (α), appropriately named Gemma, meaning "jewel," is of the second magnitude. Making a triangle with Delta (δ) and Gamma (γ) is R. This is a highly unusual variable star, which stays at about the sixth magnitude for most of the time and is thus easily seen with binoculars. However, it may suddenly fade within a few weeks to less than the tenth magnitude and disappear from binocular view. It may remain dim for a few weeks or sometimes several months. Astronomers think that this happens because the star periodically blasts off clouds of sooty matter, which blots out its light from our view.

HERCULES

Hercules is a sprawling northern constellation, covering the fifth-largest area of the sky. It is named for the Greek hero Hercules (or Heracles), who, according to legend, is the strongest man ever to have lived. He was renowned for having accomplished twelve seemingly impossible tasks, which he completed to atone for having killed his wife and children while under an evil spell.

The brightest star, Alpha (α), is faintly orange in hue, and is a red giant variable. But the brightness change, between the third and fourth magnitudes, is not easy to follow and takes place at no set interval.

The finest object in Hercules is M13, the finest globular cluster in northern skies. With a magnitude of about six, it is just visible to the naked eye and easily picked up with binoculars. Telescopes reveal it as a huge congregation of stars, clustered closely together.

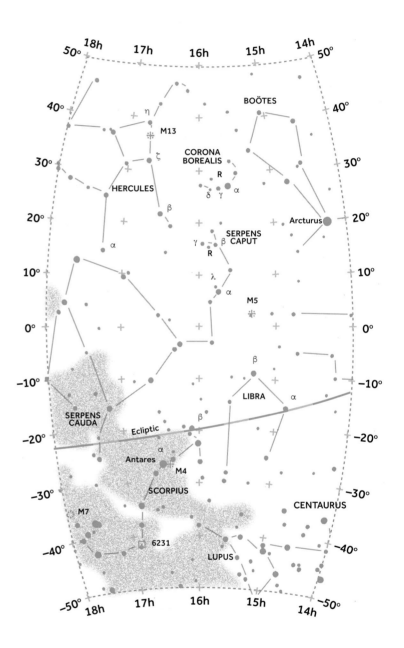

▲ Constellations visible near the meridian at about 11pm during the first week in June.

LIBRA, THE SCALES

Libra is one of the faintest constellations of the Zodiac. The Sun passes through Libra between October 31 and November 23 every year. Libra is usually associated with the scales of justice, held by the adjacent figure of Virgo. Alpha (α) is a double star, easily separated when viewed through binoculars. Beta (β) is actually slightly brighter than Alpha and has an unusual greenish tinge.

SCORPIUS, THE SCORPION

This constellation of the Zodiac is featured as the Key Constellation this month (see page 90).

▲ The globular cluster M5 in Serpens is one of the showpieces of this constellation.

▶ A cosmic light show in the Eagle Nebula, located in the Serpens constellation.

SERPENS CAPUT, THE SERPENT'S HEAD

Serpens is the only one of the 88 constellations that is split. Ophiuchus (the Serpent Bearer) divides Serpens into Caput, the head, and Cauda, the tail. The four main head stars look good through binoculars.

The constellation also boasts M5, one of the finest globular clusters in northern skies, right on the limit of naked-eye visibility. It is best found by extending a line south through Lambda (λ) and Alpha (α).

Scorpius, The Scorpion

This constellation's bright stars outline the spread claws, body, and wickedly curved tail of a scorpion poised ready to strike with its deadly sting. In Greek mythology the Scorpion stung Orion to death. In the heavens, the two constellations can never be seen together—Orion sets as Scorpius rises.

Scorpius is one of the constellations of the Zodiac, called Scorpio by astrologers, incidentally. The Sun passes through Scorpius between November 23 and 29 every year.

RIVAL OF MARS

Outshining all the other stars in Scorpius is Antares. Its name means "rival of Mars" because it is a distinctly red star, with a similar hue to the "Red Planet." Antares is also surrounded by a great red cloud, which shows up in long-exposure photos. This star is a supergiant, hundreds of times bigger than the Sun. Like many huge stars, it is unstable and pulsates, brightening and dimming over a period of about five years.

Scorpius is embedded in one of the richest regions of the Milky Way. Scanning the constellation with binoculars is highly rewarding, offering a veritable feast of star clouds and clusters, and glowing nebulae. This region is so full of astronomical delights because it is close to the center of our galaxy in neighboring Sagittarius.

Many of the deep-sky objects are visible to the naked eye and stunning in binoculars. North of the tail are two other clusters, M6 and M7. Binoculars will reveal that M6 has stars arranged like an insect with open wings, which has earned it the name of the Butterfly Cluster. Of about the fourth magnitude, M6 is easily seen with the naked eye. So is the third magnitude, M7, a broader cluster standing out against the Milky Way.

Brightest of the fine globular clusters in the constellation is M4, close to Antares. It is just on the limit of naked-eye visibility but can be difficult to make out because of the glare of Antares. Binoculars will find it straight away because it is in the same field of view as Antares.

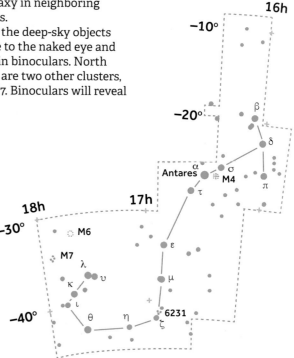

◄ NGC6357 is a diffuse nebula in the constellation Scorpius. Pismis 24 lies at its core.

▼ Many white dwarfs have been found in the M4 globular cluster in Scorpius.

June Stars

TOP: LOOKING SOUTH

Constellations visible in North America and Europe at about 11pm on or near June 7

With faint Ophiuchus moving toward the meridian, flanked by the serpent's head and tail, much of the sky is bare. Ringing this area is a curve of bright stars, beginning with Vega high in the east, followed to the west by Altair, Antares, Spica, and, high above Spica, Arcturus. Leo is close to setting beneath the western horizon. But another southern delight, Sagittarius, has just risen in the southeast.

BOTTOM: LOOKING NORTH

Constellations visible in North America and Europe at about 11pm on or near June 7

The Big Dipper (Plough) comes into our view in the west, while Cassiopeia continues to climb in the east. Deneb and the other two stars of the Summer Triangle, Altair and Vega, are climbing high too. Meanwhile, Andromeda and Pegasus have appeared in the east. But the faint misty patch of the Andromeda Galaxy may be difficult to make out in the light summer skies. Observers should have no trouble spotting the four bright stars strung roughly in line in the west—Regulus, Pollux, Castor, and Capella.

NORTHERN HEMISPHERE

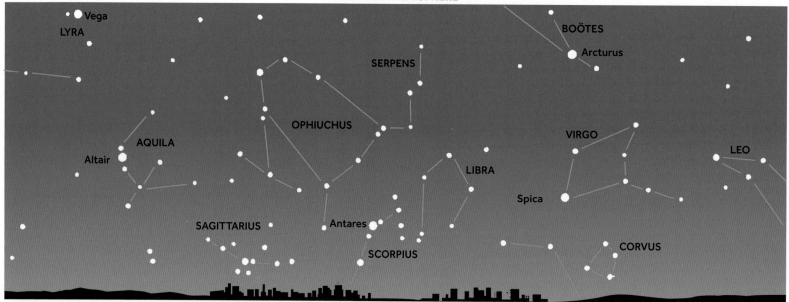

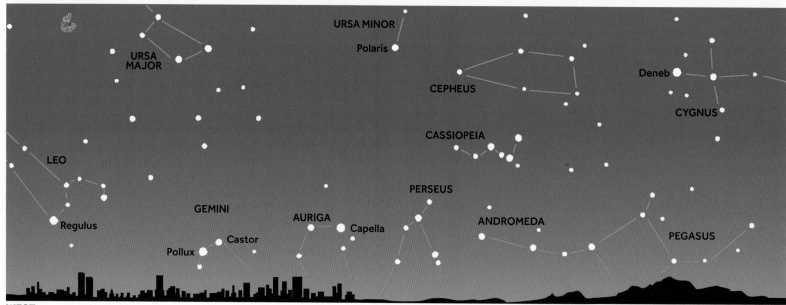

WEST   NORTH   EAST

TOP: LOOKING SOUTH

Constellations visible in Australia and South Africa at about 11pm on or near June 7

The aptly named Southern Triangle (Triangulum Australe) sits on the meridian this month. The brightest part of the sky ahead is in the southwest, where Centaurus and Crux have reappeared. Carina and Vela are still evident lower down, with Canopus close to the horizon. The skies of the southeast are dominated by three of the southern birds—the Peacock (Pavo), Toucan (Tucana), and Crane (Grus). The Southern Fish (Piscis Austrinus), with lead star Fomalhaut, is just rising—perhaps unwisely—beneath the crane.

BOTTOM: LOOKING NORTH

Constellations visible in Australia and South Africa at about 11pm on or near June 7

With faint Ophiuchus near the meridian, flanked by the serpent's head and tail, much of the sky is bare. Ringing this area is a curve of bright stars, beginning with Spica high in the west, followed to the east by Arcturus, Vega (close to the horizon), and Altair. The familiar northern constellations Boötes and Hercules appear on either side of the meridian, with the curve of Corona Borealis between them.

SOUTHERN HEMISPHERE

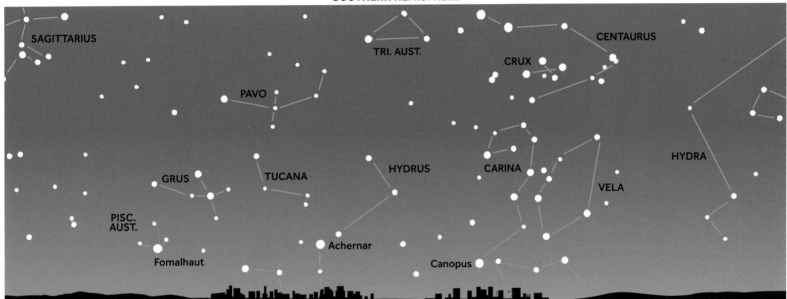

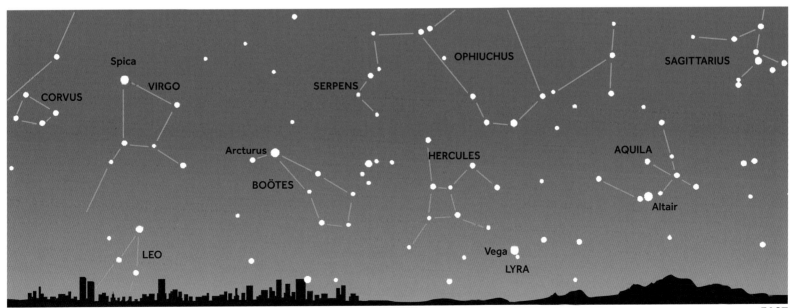

WEST   NORTH   EAST

July Stars

This month shows three stellar beacons that herald summer—Deneb in Cygnus, Vega in Lyra, and Altair in Aquila—which form the conspicuous Summer Triangle. The name can be used by observers in the Southern Hemisphere, although there, of course, it is now winter! The southern skies this month are dazzling, with Scorpius still prominent and Sagittarius in hot pursuit as the heavens revolve. Northern observers, alas, can catch only a glimpse of these magnificent constellations.

LYRA, THE LYRE

The lyre was played by the most celebrated musician in Greek mythology, Orpheus. Lyra is only a tiny constellation, but it is full of interest. It has only one bright star, Vega, which is the fifth-brightest star in the heavens. It is sometimes called the Harp Star, but its name means "swooping eagle" in Arabic, for the Arabs saw the constellation as an eagle. Historically, Vega was the Pole Star in about 10,000 BC and will become so again in AD 14,500. This is the result of the Earth's axis gradually changing direction in space, a movement known as "precession."

Close to Vega is a star that is a particular favorite among northern astronomers. It is the "double-double" Epsilon (ε). The very sharp-sighted may be able to spot that it is a double star with the naked eye, and the pair of stars are easily separated through binoculars. Seen through a small telescope, each star of the pair can be seen to be a double too.

Beta (β) is another fascinating star. It is made up of a small bright star and a large dimmer star. The two revolve around each other, and every 13 days the dim one covers the bright one, causing Beta to fade briefly.

OPHIUCHUS, THE SERPENT BEARER

This is a large constellation depicting a man carrying a snake. Since the snake sheds its skin every year, it has long been a symbol for renovation and healing. As a constellation,

Ophiuchus splits the snake (Serpens) in two—Caput (Head) and Cauda (Tail). Its stars are not particularly impressive, with Alpha (α) being only of the second magnitude.

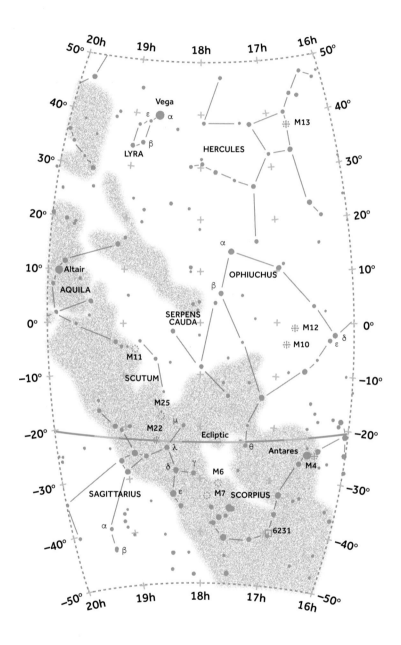

▲ Constellations visible near the meridian at about 11pm during the first week in July.

SAGITTARIUS, THE ARCHER

The Sun passes through this constellation of the Zodiac between December 18 and January 19 every year. It is one of the most spectacular of all the constellations, embedded in the richest region of the Milky Way. The center of our galaxy lies in this direction.

The ancient Greeks thought of the figure as Crotus, who invented archery and was the son of the pipe-playing god Pan. He is depicted as a centaur—half-man, half-horse. He is pulling a bow, with the arrow aimed at Antares, the heart of the Scorpion.

The group of five stars in the center of the constellation is often called the Milk Dipper after its shape and the fact that it "dips" into the Milky Way. Add Gamma (γ), Epsilon (ε), and Delta (δ) to the group, and it becomes the Teapot, with a pointed lid and long spout.

The individual stars in the constellation are not particularly interesting, although Beta (β) is a naked-eye and binocular double. It is in the surrounding Milky Way that Sagittarius is so spectacular. There is a host of globular clusters, none more magnificent than M22, the third-brightest globular in the heavens and visible to the naked eye. The region west of the stars Lambda (λ) and Mu (μ) is especially rich in clusters and nebulae, which include the Lagoon (M8) and Trifid (M20).

SERPENS CAUDA, THE SERPENT'S TAIL

The tail end of the divided constellation Serpens is much less impressive than the head (Serpens Caput, see page 89). Its southern end dips into the Milky Way and in this region, just west of Gamma (γ) in neighboring Scutum, is the delightful Eagle Nebula (M16). This is the site of one of the Hubble Space Telescope's most dramatic photos, named The Pillars of Creation (see following pages).

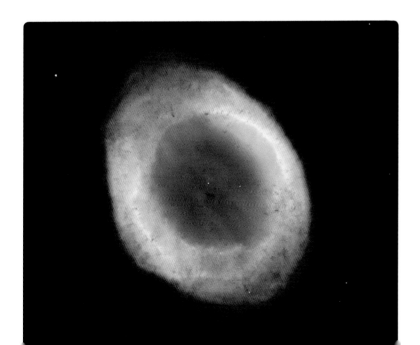

◄ One of Lyra's best-known features, the planetary nebula that looks just like a smoke ring. It is named the Ring Nebula.

▲ Clouds in the Eagle Nebula, where stars are being born.

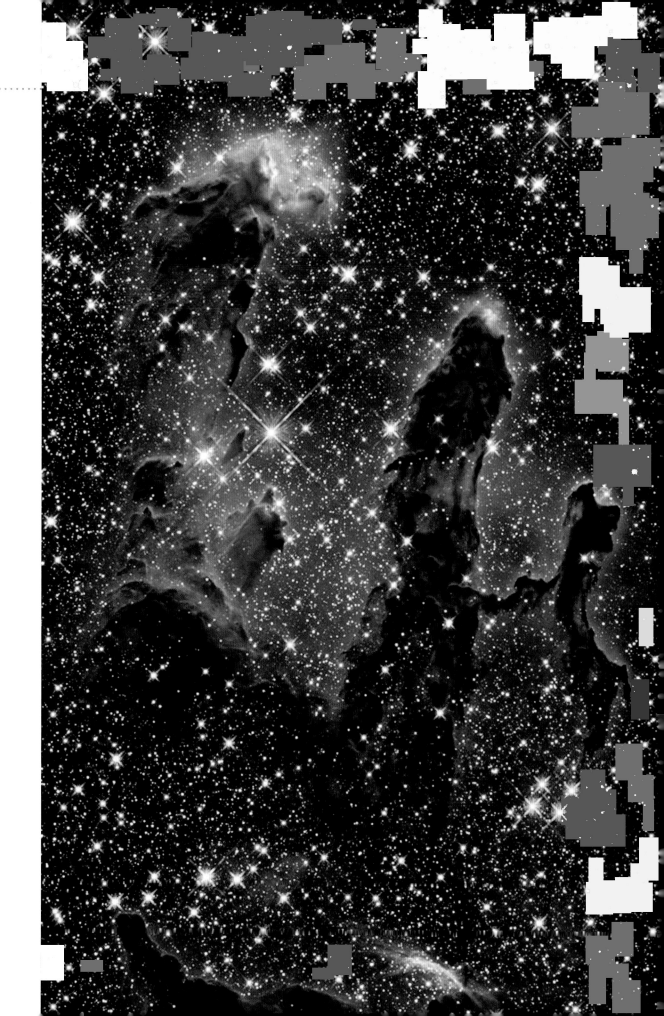

◀ NASA's Hubble Space Telescope first captured this stunning view of the so-called "Pillars of Creation" in 1995. It shows three vast columns of gas illuminated by ultraviolet light from a cluster of young stars in the Eagle Nebula, or M16.

▶ Returning to the site of its most iconic image, the NASA/ESA Hubble Space Telescope captured the pillars again. As seen in infrared light, it reveals the famous pillars surrounded by stars in the process of being formed.

July Stars

TOP: LOOKING SOUTH

Constellations visible in North America and Europe at about 11pm on or near July 7

July is another good month to peer deep into the southern hemisphere. This is the month when observers can see Sagittarius furthest above the horizon, although Scorpius is beginning to set. Arcturus and Spica (low down) form a bright pair in the west. Aquila is coming up to the meridian, with Altair brilliant. With Deneb in Cygnus and Vega in Lyra overhead (and out of the frame here), Altair forms the prominent Summer Triangle.

BOTTOM: LOOKING NORTH

Constellations visible in North America and Europe at about 11pm on or near July 7

Both bears, Ursa Major and Minor, are now descending. Regulus has disappeared beneath the western horizon and the rest of the constellation of Leo will soon be following. Capella is conspicuous, now low down on the northern horizon as Boötes straddles the meridian. Cassiopeia still delights in mid-skies to the east, and the starry background of the Milky Way provides a feast in binoculars. Andromeda is better placed now for observation, and the Square of Pegasus has fully risen.

NORTHERN HEMISPHERE

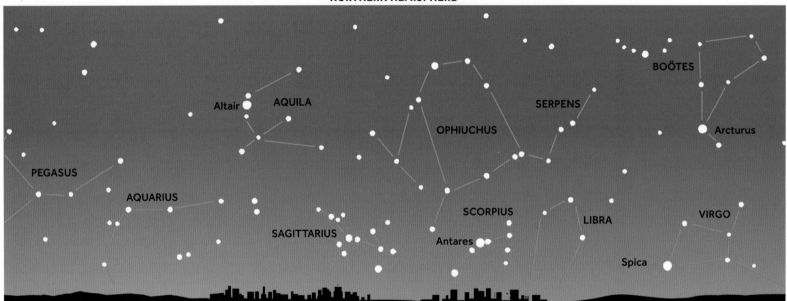

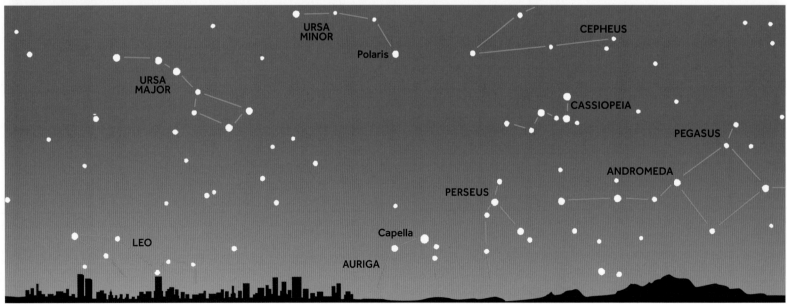

TOP: LOOKING SOUTH

Constellations visible in Australia and South Africa at about 11pm on or near July 7

The dazzling Milky Way stands nearly vertical this month just west of the meridian, running from Scorpius high overhead (out of the frame here) through Carina and Vela, now sinking beneath the horizon. Canopus has disappeared, while Achernar is now climbing in the southeast. The only other bright star in eastern skies is Fomalhaut.

BOTTOM: LOOKING NORTH

Constellations visible in Australia and South Africa at about 11pm on or near July 7

Arcturus and Spica are now low in the west and will shortly disappear, as will Hercules. Above them the skies lack interest, occupied mainly by Ophiuchus and Serpens. Vega in Lyra is close to the meridian. Deneb is just appearing over the horizon further east, while higher up is Altair. These three stars make up the so-called Summer Triangle—oddly named for southern observers due to its appearance in mid-winter.

SOUTHERN HEMISPHERE

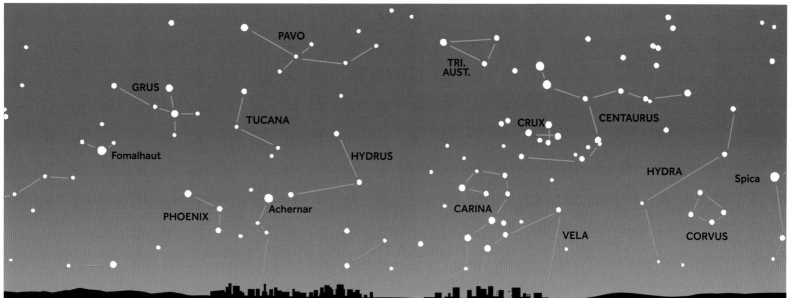

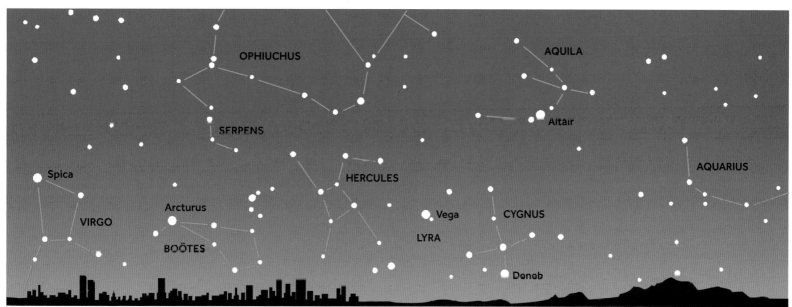

WEST · NORTH · EAST

August Stars

The Summer Triangle still blazes in northern skies, this month straddling the meridian in late evening. On moonless nights, the Milky Way glows overhead, bisecting the sky.

AQUILA, THE EAGLE

This prominent bird wings its way through the Milky Way just south of Cygnus, straddling the celestial equator. In mythology, Aquila was the favorite bird of Zeus, and was used to retrieve the thunderbolts Zeus hurled at his enemies.

Aquila's leading star, Altair, is the southernmost of the trio of stars that form the Summer Triangle, along with Deneb in Cygnus and Vega in Lyra. It is one of the closest of the bright stars, being about 17 light-years away.

CAPRICORNUS, THE SEA GOAT

This small constellation of the Zodiac is composed of faint stars roughly in the shape of a rather crooked triangle. The Sun passes through it between January 19 and February 16. Capricornus is depicted as a fish-

tailed goat, associated with the goat-headed god Pan.

Both Alpha (α) and Beta (β) are doubles visible with binoculars, but aside from these there is not a great deal of interest in the constellation.

CYGNUS, THE SWAN

The distinctive northern grouping of Cygnus is the Key Constellation this month (see page 102).

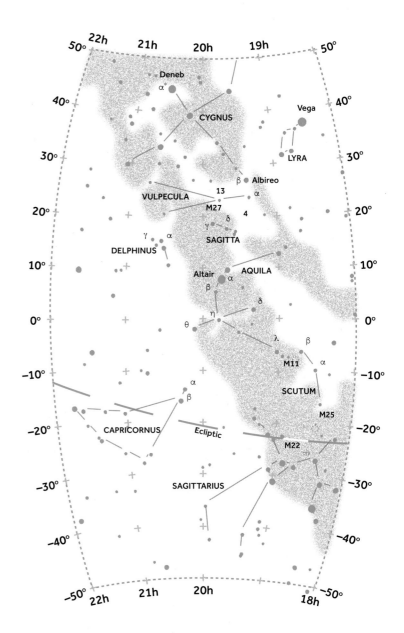

▲ Constellations visible near the meridian at about 11pm during the first week in August.

DELPHINUS, THE DOLPHIN

Delphinus is a tiny constellation, but one that needs only a little imagination to picture as a dolphin leaping gracefully through the waves. In Greek mythology, the dolphin was immortalized in the heavens for its help in bringing one of the beautiful sea nymphs, the Nereids, to be the wife of the sea god, Poseidon. The distinctive parallelogram of four stars in Delphinus looks almost like a star cluster. Gamma (γ) is a fine double for small telescopes.

SAGITTA, THE ARROW

Another tiny constellation, Sagitta does indeed look rather arrowlike. Being in the Milky Way, Sagitta is worth scanning with binoculars, for it has some really nice star fields. Binoculars and small telescopes will pick up a globular cluster (M71) midway between Gamma (γ) and Delta (δ).

SCUTUM, THE SHIELD

This small constellation is reasonably modern (1600s) and has no mythogical connections. Its main item of interest is a rich open cluster, M11. It is visible through binoculars as a fuzzy ball, which a small telescope resolves into a V-shaped formation of stars.

▶ Top: The Milky Way brightens August skies north and south. This is the Lagoon Nebula in Sagittarius.

▶ Bottom: A "hole" in Cygnus is caused by a dark cloud of gas and dust.

VULPECULA, THE FOX

Vulpecula is a modern constellation (1600s) of faint stars south of Cygnus. Being in the Milky Way, it contains some fine starscapes for binocular viewers. Just south of Star 13 and north of Gamma (γ) in neighboring Sagitta, is the planetary nebula, M27. Larger telescopes show off its distinctive shape that earns it the name of the Dumbbell Nebula. Star 4, due south of Alpha (α), is set in a charming pattern of about ten stars that go by the name of the Coat Hanger.

Cygnus

Little imagination is required to picture these stars as a swan winging its way across the northern sky, as if flying south on migration. Its curved wings are widespread, poised for a powerful downbeat; its long neck is outstretched, its tail fanned out.

In Greek mythology, the swan was one of the many disguises adopted by Zeus when he went about his illicit affairs with fair maidens. The result of his tryst with Leda, Queen of Sparta, was that Leda laid eggs, out of which hatched the twins Castor and Pollux (which feature in the constellation Gemini), and the legendary beauty Helen of Troy.

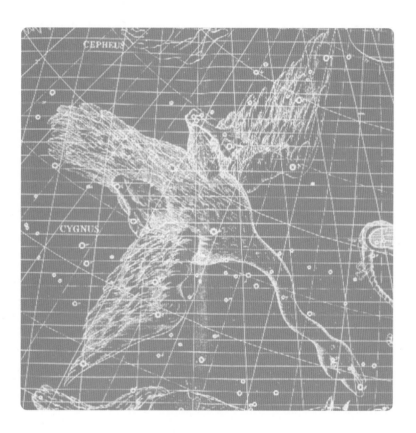

DELIGHTS GALORE

Brightest of the stars in the constellation is Deneb, in the swan's tail, which is a blue-white supergiant. It lies around 2,600 light-years from us, and to appear so bright at such a distance it must have the energy output of more than 60,000 suns. Albireo, at the opposite end of the swan figure, is much fainter, but it is a fine double star. Small telescopes will separate it into lovely blue and yellow components.

Cygnus is among the richest of all the constellations in the northern sky. Close to Deneb, telescopes show a glowing red nebula (NGC7000) that bears an uncanny resemblance to the continent of North America. South of Epsilon (ε), long-exposure photographs reveal a vast loop of glowing gas, the Cygnus Loop, which is all that remains of a gigantic star that blasted itself to pieces.

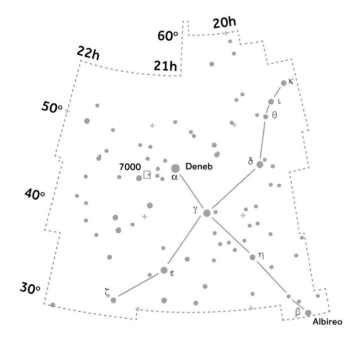

▶ Facing page: Part of the Cygnus Loop, a still-expanding cloud from an ancient supernova explosion.

August Stars

TOP: LOOKING SOUTH

Constellations visible in North America and Europe at about 11pm on or near August 7

Aquila now straddles the meridian, with Altair nearly due south. The Summer Triangle it makes with Deneb and Vega (out of the frame) is almost directly overhead. This is the last month to enjoy the brilliance of Sagittarius before it slips below the southern horizon.

BOTTOM: LOOKING NORTH

Constellations visible in North America and Europe at about 11pm on or near August 7

Arcturus, which we last spotted in eastern skies in February, now makes an appearance in the west. Arcturus, at the tail end of the kite-shaped Boötes, is the fourth brightest in the whole heavens. Its appearance this month is welcome, because in this view the only other bright star is Capella, now climbing. High overhead, however, the Summer Triangle is still with us.

NORTHERN HEMISPHERE

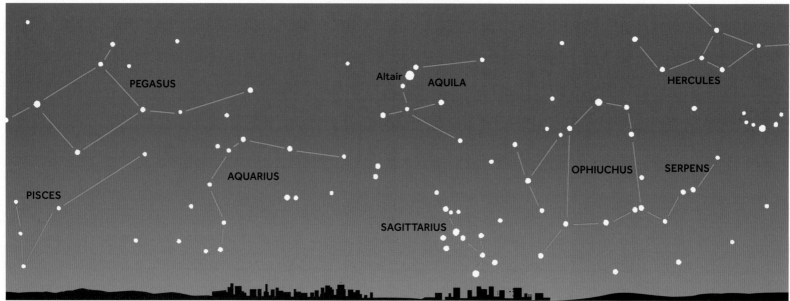

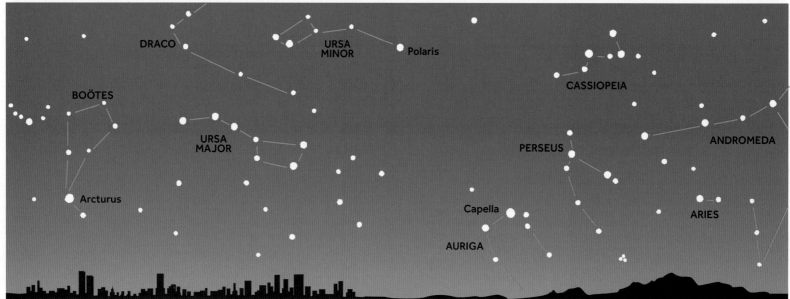

WEST　　　　　　　　　　NORTH　　　　　　　　　　EAST

TOP: LOOKING SOUTH

Constellations visible in Australia and South Africa at about 11pm on or near August 7

This month all the spectacle in the southern heavens is concentrated in the southwest around the Milky Way, where Scorpius can be seen again. But the rest of the sky is dull by contrast, being occupied by a flock of birds—the Phoenix, Toucan (Tucana), Crane (Grus), and Peacock (Pavo), which is now on the meridian. For the second month, only Achernar and Fomalhaut brighten up the east.

BOTTOM: LOOKING NORTH

Constellations visible in Australia and South Africa at about 11pm on or near August 7

The Summer Triangle sits on the meridian this month, with Vega and Deneb low down and Vega in mid-skies. The only other really bright stars on show are Antares in the far west and Fomalhaut in the far east (out of the frame). Low in the east, Pegasus has risen, its famous square empty of bright stars. It is sometimes called the Fall (or Autumn) Square, but this again is relevant only to northern observers, because spring is approaching in southern skies.

SOUTHERN HEMISPHERE

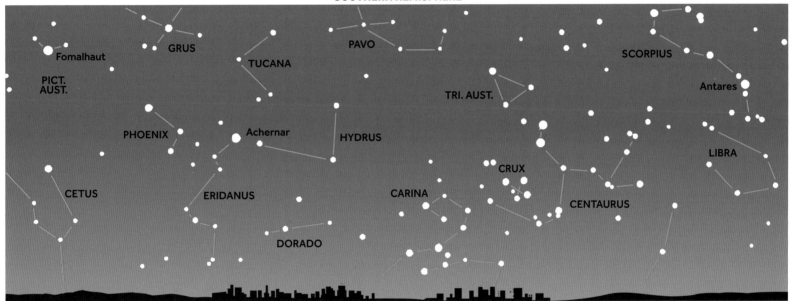

Fomalhaut · GRUS · PAVO · SCORPIUS
PICT. AUST. · TUCANA · TRI. AUST. · Antares
PHOENIX · Achernar · HYDRUS · LIBRA
CETUS · ERIDANUS · CRUX · CENTAURUS · CARINA
DORADO

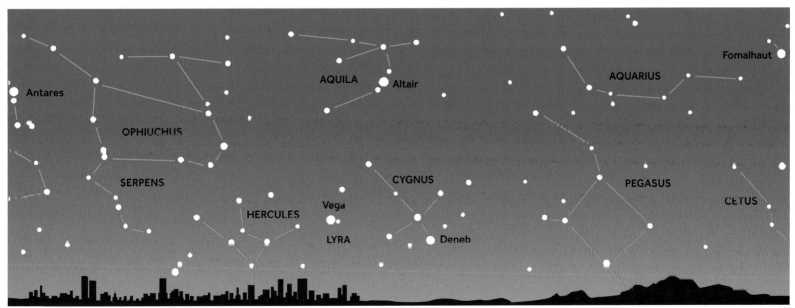

Antares · AQUILA · Altair · AQUARIUS · Fomalhaut
OPHIUCHUS · SERPENS · CYGNUS · PEGASUS · CETUS
HERCULES · Vega · LYRA · Deneb

WEST **NORTH** **EAST**

September Stars

The Milky Way and its necklace of glittering constellations—Cygnus, Aquila, and Sagittarius—is disappearing fast. The skies become bare again except for Pegasus.

AQUARIUS, THE WATER BEARER

The Sun passes through this Zodiac constellation between February 16 and March 11. It is a large constellation, though not particularly easy to make out.

Aquarius depicts the figure of a youth pouring water from a jar. It pours out as a cascade of stars and ends up in the mouth of Piscis Austrinus, the Southern Fish. The figure is usually identified with the beautiful boy Ganymede, whom Zeus took back to Mount Olympus to dispense wine to the gods.

The mouth of the jar, out of which the water pours, is marked by the four stars Eta (η), Pi (π), Gamma (γ), and Zeta (ζ). Zeta is a double star, visible in telescopes. The globular cluster M2 is due west of Alpha (α) and due north of Beta (β). It is too faint to be visible to the naked eye but is easily seen in binoculars and a small telescope.

The nebula NGC7009 can be spotted in small telescopes just south of Epsilon (ε). It is a planetary nebula, so called because it shows up as a disk, rather like a planet. This particular planetary nebula looks like a ringed planet, and is called the Saturn Nebula.

GRUS, THE CRANE

Grus is a relatively modern (1600s) constellation that was, for many years, referred to as the Flamingo. It is easily located immediately south of the bright Fomalhaut in Piscis Austrinus.

Its two brightest stars, Alpha (α) and Beta (β), contrast nicely, since Alpha is brilliant white, while Beta is orange. Delta (δ) is a double star that can be separated by the naked eye; it is made up of yellow and red giant stars.

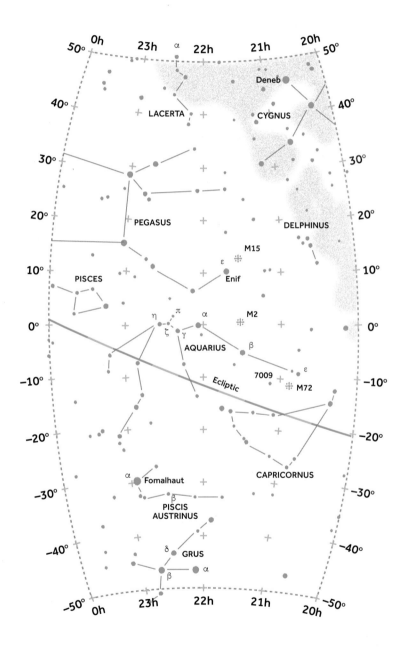

▲ Constellations visible near the meridian at about 11pm during the first week in September.

▲ Above top: Comet Hale-Bopp moves through Lacerta on March 9, 1997.

▲ Above: Little blobs of gas with tails form at the edge of the Helix Nebula in Aquarius. Astronomers call them "cometary knots."

▶ Right: The barrel spiral galaxy IC5201 in the constellation Grus.

LACERTA, THE LIZARD

Lacerta is a small northern constellation made up of a zigzag of faint stars. Its brighter head stars lie in the Milky Way, which, as always, is worth a sweep with binoculars. Quite a bright cluster shows up just to the west of Alpha (α).

PISCIS AUSTRINUS, THE SOUTHERN FISH

Gulping water poured by Aquarius, the Southern Fish was considered to be the parent of the fishes that formed the Zodiacal constellation Pisces. It is a small constellation, lifted from obscurity by its first-magnitude lead star Fomalhaut, meaning "fish's mouth" in Arabic.

September Stars

TOP: LOOKING SOUTH

Constellations visible in North America and Europe at about 11pm on or near September 7

The flying horse Pegasus continues its way west. With Aquarius in mid-skies on the meridian, Pisces swimming behind, and Cetus rising, the heavens have taken on a watery aspect. This is reinforced by the rising almost due south of Fomalhaut and the Southern Fish.

BOTTOM: LOOKING NORTH

Constellations visible in North America and Europe at about 11pm on or near September 7

This month the Big Dipper, or Plough, reaches its lowest point, with the handle nearly horizontal. The two stars that form the pointers to Polaris, Merak (lower) and Dubhe, are close to the meridian and pointing almost vertically upward. To the west, Arcturus is close to setting and may be difficult to spot. In the east, however, Capella has been joined by the red eye of the bull, Aldebaran, as Taurus rises over the horizon. The Pleiades, located almost vertically above Aldebaran, are unmistakable.

NORTHERN HEMISPHERE

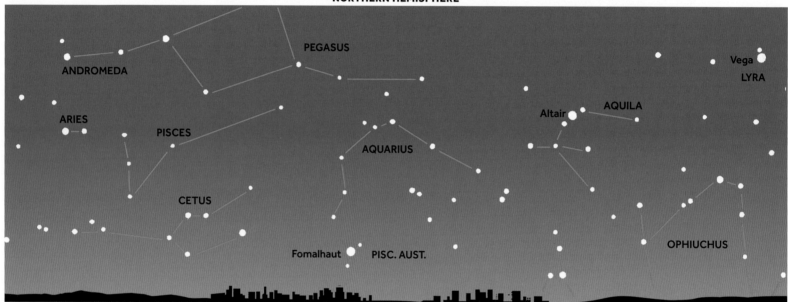

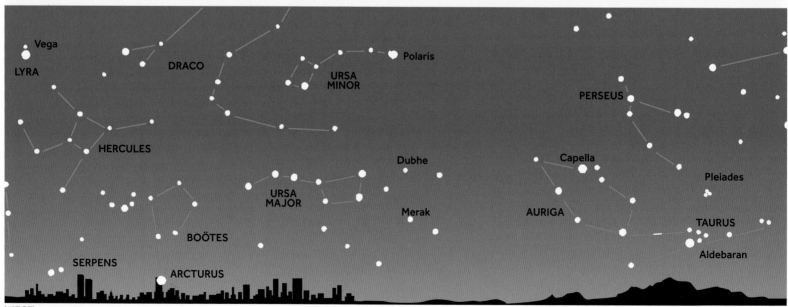

WEST　　　　　　NORTH　　　　　　EAST

TOP: LOOKING SOUTH

Constellations visible in Australia and South Africa at about 11pm on or near September 7

Southern skies brighten this month. Canopus has reappeared in the southeast and is climbing. Its constellation—Carina—has drawn clear of the horizon. Centaurus, Scorpius, and Sagittarius all dazzle in the west. The east looks boring by comparison, occupied mainly by the widely meandering river Eridanus. The southern birds—the phoenix, toucan, crane, and peacock—are now flying high.

BOTTOM: LOOKING NORTH

Constellations visible in Australia and South Africa at about 11pm on or near September 7

The flying horse Pegasus continues winging its way west across the heavens. Aquarius is in mid-skies, with Pisces following behind, and Cetus is coming into view. The northeast skies remain on the dull side; most interest is in the northwest, with the two birds, the Eagle and the Swan, conspicuous and nicely placed for observation.

SOUTHERN HEMISPHERE

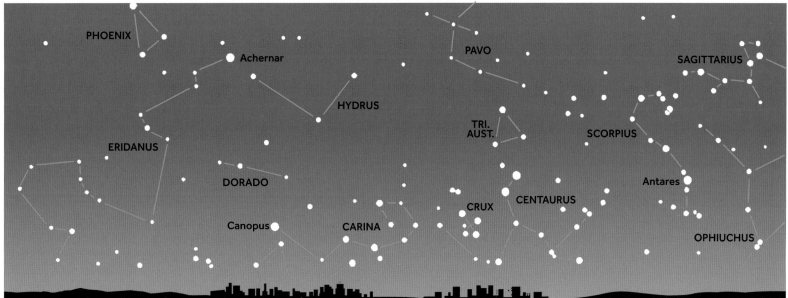

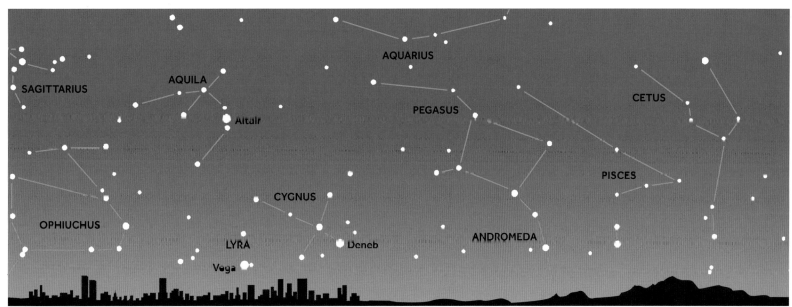

October Stars

This month the Square of Pegasus sits right on the meridian and reminds us that fall in the Northern Hemisphere and spring in the Southern are well advanced. Few other constellations immediately strike the eye, as their stars are generally faint. Only in the far south does Fomalhaut continue to shine like a beacon.

PEGASUS, THE FLYING HORSE

One of the most ancient of the constellations, Pegasus has always been associated with the winged horse. The winged horse was a favorite theme among the artists of Assyria, one of the earliest Mesopotamian civilizations. In Greek myths, Pegasus was the winged horse that leaped out of the corpse of the monstrous Medusa after she had been beheaded by the hero Perseus.

The Square of Pegasus, the near perfect square made by the stars Beta (β), Alpha (α), Gamma (γ), and Alpha in Andromeda, is one of the most recognizable star patterns in the heavens. All four stars are of about the same (second) magnitude. Of equal brightness is Epsilon (ε), just out of the frame here, but shown on the previous map (page 106). Named Enif, it represents the horse's nose. A short distance away from Enif is the fine globular cluster M15. Visible in binoculars, it has a highly luminous core, which strongly emits X-rays, suggesting there is a black hole lurking within.

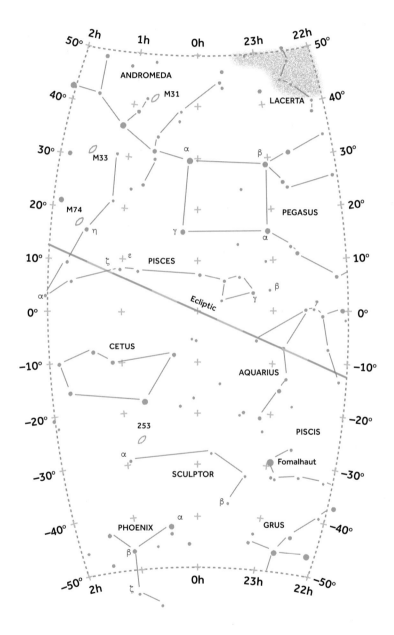

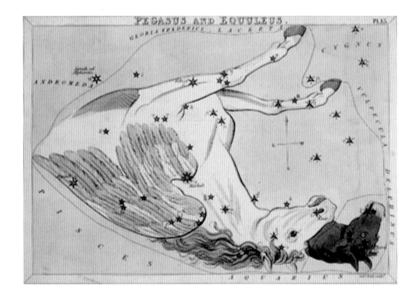

▲ A fine portrayal of Pegasus flying across the heavens.

▲ Constellations visible near the meridian at about 11pm during the first week in October.

PHOENIX, THE PHOENIX

A relatively modern constellation (1500s), the Phoenix is named after the fabled bird that was reincarnated from its own ashes. After living for 500 years, it burned itself on a funeral pyre, only to fly out of the flames reborn. Zeta is a double that is easy to separate. Its brighter component varies in brightness every 40 hours, since it is an eclipsing binary.

PISCES, THE FISHES

The Sun passes through large but faint Pisces between March 12 and April 18 every year. In mythology, the two fishes represent Venus and her son Cupid. They had to dive into the River Euphrates to escape from the monster Typhon.

Pisces is the constellation in which the ecliptic and celestial equator intersect, or, in other words, where the Sun appears to cross the celestial equator as it travels north. This happens on about March 21 every year. This is the time of the spring, or vernal equinox, which marks the start of spring in the Northern Hemisphere and of fall in the Southern.

The intersection of the ecliptic and the celestial equator is also the starting point (0 hours) for celestial longitude, or right ascension (see page 52).

SCULPTOR, THE SCULPTOR

Sculptor is yet another faint constellation in this part of the sky, found by scanning east from the bright Fomalhaut in the neighboring Piscis Austrinus. This is a modern constellation, introduced in the 1700s, and has no mythological connections. Its main attraction is NGC253 to the north of Alpha (α).

◀ Top: One of the Sculptor group of galaxies, this is NGC253, a spiral galaxy that we see sideways. Our own galaxy would look much like this from a distance.

◀ Bottom: UGC477, a diffuse, low-surface brightness galaxy in the constellation Pisces.

October Stars

TOP: LOOKING SOUTH

Constellations visible in North America and Europe at about 11pm on or near October 7

This month the Square of Pegasus sits on the meridian. The skies elsewhere are far from brilliant, occupied by Pegasus, Andromeda (just out of the frame), Pisces, Aries, Aquarius, Cetus, and the river Eridanus, which has appeared over the southeast horizon. But, as the weather cools and skies get darker, we see Taurus risen and Orion peeping over the horizon.

BOTTOM: LOOKING NORTH

Constellations visible in North America and Europe at about 11pm on or near October 7

In the west, the Summer Triangle of Deneb, Vega, and Altair is making its descent. In the east, Gemini has risen, with its two lead stars Castor (top) and Pollux one above the other. At the same height as Pollux and further east, the noticeably red Betelgeuse puts in an appearance as Orion begins to rise.

NORTHERN HEMISPHERE

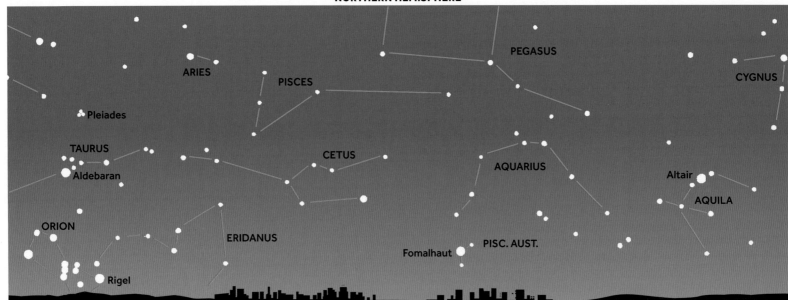

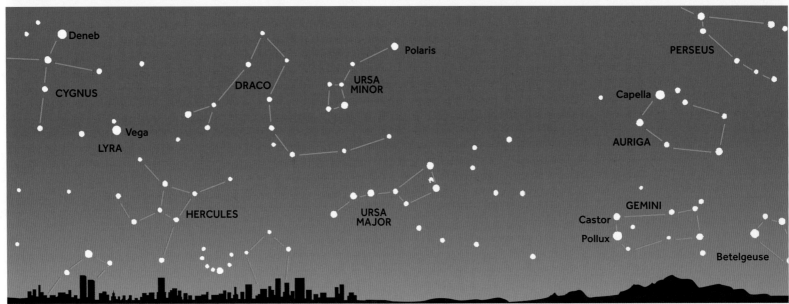

WEST NORTH EAST

TOP: LOOKING SOUTH

Constellations visible in Australia and South Africa at about 11pm on or near October 7

From the horizon up to mid-skies, the heavens this month are outstanding. Orion is rising in the east, with Rigel well above the horizon. Canis Major has mostly risen too, with Sirius climbing into view. Canopus leads the march of the three nautical constellations—keel (Carina), poop deck (Puppis), and sails (Vela). Crux is close to the meridian low down.

BOTTOM: LOOKING NORTH

Constellations visible in Australia and South Africa at about 11pm on or near October 7

This month the Square of Pegasus sits on the meridian. The northern aspects of the skies, however, are far from brilliant, with most of the heavens occupied by relatively faint constellations—Pegasus, Andromeda, Pisces, Aries, Aquarius, Cetus, and Eridanus. This month is a good time to observe the Andromeda Galaxy, close to the meridian.

SOUTHERN HEMISPHERE

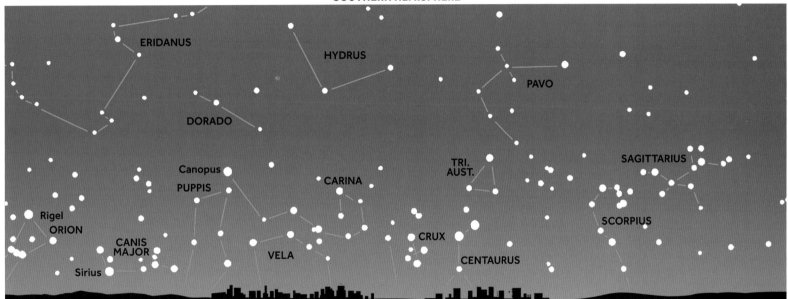

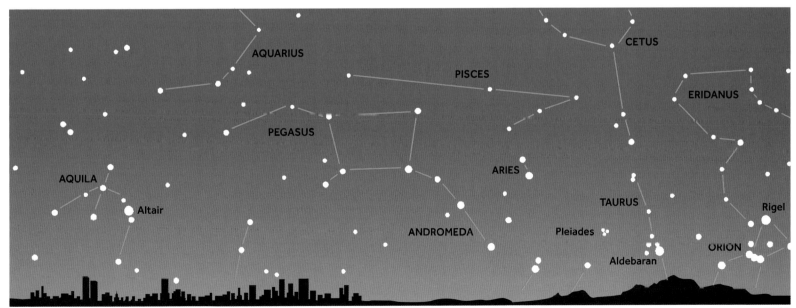

WEST　　　　　　　　　　　　　　　　　NORTH　　　　　　　　　　　　　　　　　EAST

November Stars

Meridian skies are far from spectacular this month. Pegasus has slipped away west, and Taurus and Orion are still far in the east. Andromeda, however, provides a spectacular highlight, revealing as a faint misty patch the galaxy that represents the furthest object in the heavens the naked eye can see, at a distance of 24 quintillion km.

ANDROMEDA

While the pattern of stars making up this constellation is not particularly memorable, Andromeda can easily be found because it is joined to one of the most distinctive star groups in the sky—the Square of Pegasus. Andromeda was the lovely daughter of King Cepheus and Queen Cassiopeia. The sea god Poseidon became upset when Cassiopeia boasted about how beautiful she was. To placate the god, Andromeda was chained to a rock to provide a meal for the sea monster Cetus. All ended happily, however, when the hero Perseus happened by and snatched the fair maiden from the monster's jaws in the nick of time.

Perhaps the most appealing of the constellation's stars is Gamma (γ), which is an easy double for small telescopes. The two components make a lovely pair, the one bright orange, the other blue-green. But it is the fuzzy patch seen in the north of Andromeda that merits the most attention. Once thought to be a gas cloud, it was named the Great Nebula in Andromeda. However, it is actually another galaxy, cataloged as M31, lying 2.5 million light-years away.

The Andromeda Galaxy is one of the few that can be seen with the naked eye, but binoculars or a small telescope show it better. The Andromeda Galaxy is one and a half times as big as our own galaxy and is in a cluster we call the Local Group.

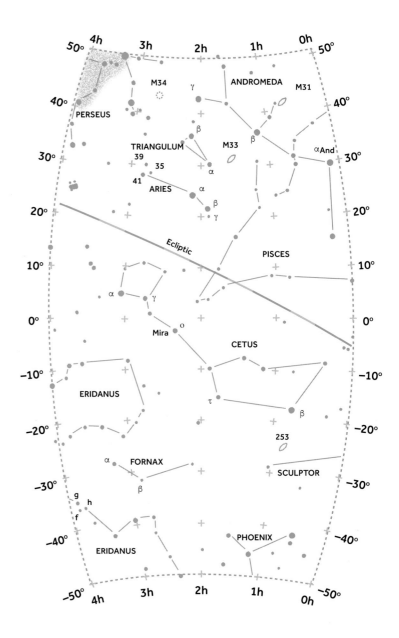

▲ Constellations visible near the meridian at about 11pm during the first week in November.

ARIES, THE RAM

Aries is one of the smaller constellations of the Zodiac, sandwiched between Taurus and Pisces. The Sun travels through Aries between April 18 and May 14 every year. Aries represents the ram with the Golden Fleece sought by Jason and the Argonauts.

It is best located by reference to the Square of Pegasus in the west and the Pleiades in the east. Gamma (γ) is a double visible in small telescopes. The three stars 35, 39 and 41 form a little triangle that was once known as Musca Borealis, or the Northern Fly, representing flies buzzing around the Ram's tail.

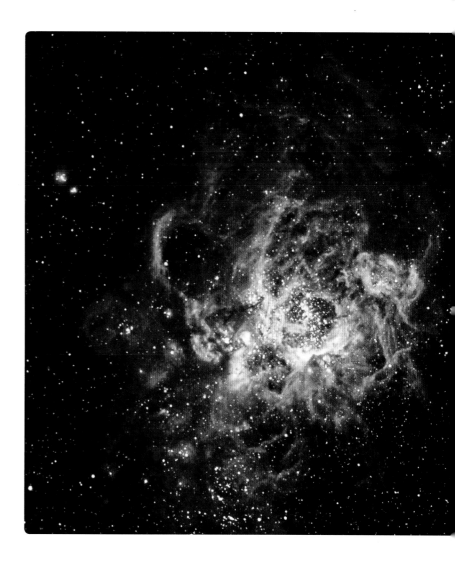

CETUS, THE WHALE

Also known as the Sea Monster, this is one of the largest constellations. It represents the monster that was about to devour the fair Andromeda (see opposite). Cetus is not easy to trace because its stars are mostly faint; even its brightest—Alpha (α) and Beta (β)—are only third magnitude. Gamma (γ) is a lovely binary for small telescopes, with blue and yellow parts. Tau (τ) is of interest because it is a yellow dwarf star that is nearly identical to the Sun. But the pièce de résistance of the constellation is the star Omicron (ο), better known as Mira, meaning "the Wonderful." It got its name because it was the first red-giant variable star to be discovered, by the Dutch astronomer David Fabricus in 1596. Mira varies in brightness noticeably over a period of about 11 months. At maximum, it is easily visible to the naked eye at around the third magnitude, but it fades at minimum to about the tenth magnitude.

▲ A star-forming region in the beautiful spiral galaxy M33.

TRIANGULUM, THE TRIANGLE

This small constellation could not be better named, as its trio of main stars form a perfect right-angled triangle. The object to look for in Triangulum is M33, a spiral galaxy that presents itself to us head-on. It is just bright enough to spot with the naked eye on a really dark night, between Alpha (α) and the star Beta (β) in Andromeda.

November Stars

TOP: LOOKING SOUTH

Constellations visible in North America and Europe at about 11pm on or near November 7

While Pegasus is flying ever westward—along with the dull constellations of Pisces, Cetus, and Aquarius—Orion is striding across the heavens in the east. He faces the charging bull Taurus, with horns lowered and glaring eye, marked by Aldebaran. Slightly higher in the sky are the Pleiades, the star group named after the sisters whom Orion chased in Greek mythology. This outstanding open or galactic cluster is now well placed for observation.

BOTTOM: LOOKING NORTH

Constellations visible in North America and Europe at about 11pm on or near November 7

This month the Swan (Cygnus) is flying vertically in the west, with its long neck outstretched and wings spread wide, as if diving headlong for the horizon. However, the other bird, the Eagle (Aquila), will get there first. Their bright stars still form a conspicuous triangle with Vega (Lyra). In the east, Canis Minor has risen, and Procyon joins Castor, Pollux, and Capella.

NORTHERN HEMISPHERE

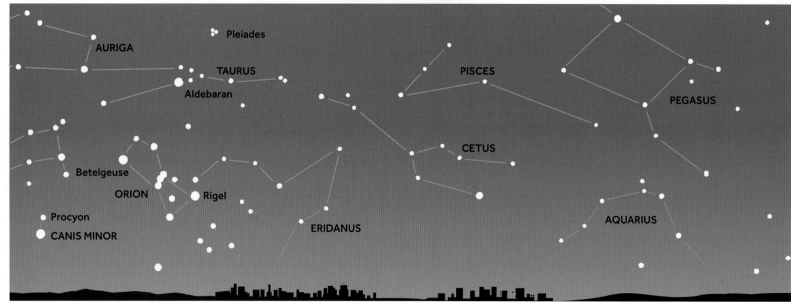

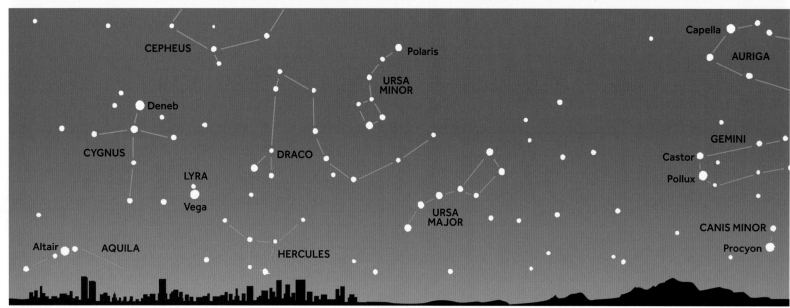

WEST NORTH EAST

TOP: LOOKING SOUTH

Constellations visible in Australia and South Africa at about 11pm on or near November 7

The lower part of the skies are not quite as dazzling this month. Scorpius and Sagittarius have mostly disappeared in the southwest, taking with them the most brilliant region of the Milky Way. The southeast is now the brightest aspect of the skies, with Sirius and Canopus relatively close. Low in the south the twin bright stars Alpha and Beta Centauri lie close to the meridian, while Crux has just passed it.

BOTTOM: LOOKING NORTH

Constellations visible in Australia and South Africa at about 11pm on or near November 7

As Pegasus continues to fly westward—along with the constellations of Pisces, Aries, and Aquarius—the stunning Orion strides across the heavens in the east to confront the charging bull Taurus, whose eye is marked by Aldebaran. In the east now are the two celestial dogs Canis Major and Minor (just out of frame, far right). Brightest-star-in-the-sky Sirius is well above the horizon, although Procyon is low down. The Andromeda Galaxy is again well placed for observation.

SOUTHERN HEMISPHERE

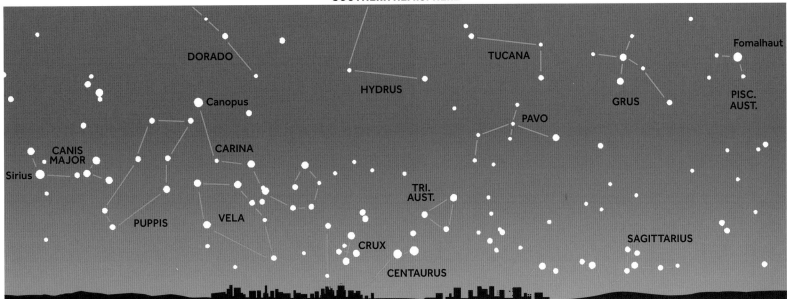

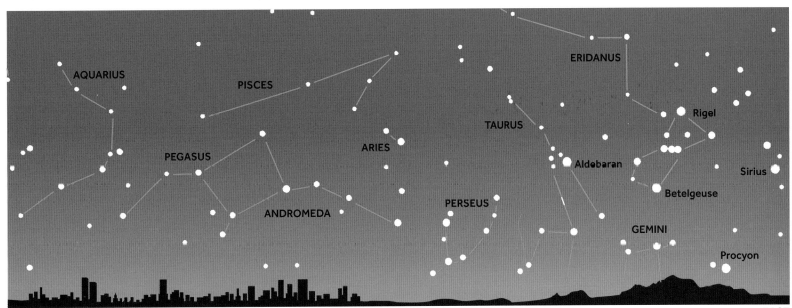

WEST · NORTH · EAST

December Stars

This month the faint stars of the past few months have all but drifted away, as dazzling constellations and the Milky Way are moving in to take their place, from Perseus in the north, through Taurus and Orion, to Canis Major in the south. The night sky's most famous cluster, the Seven Sisters, or the Pleiades, is now well placed for observation from both hemispheres.

ERIDANUS

This immensely long constellation winds itself through the southern heavens. It is no small wonder, therefore, that Eridanus has been identified with rivers from the earliest times.

Eridanus rises near the bright Rigel in Orion, and leisurely makes its way south. Its mouth is marked by its brightest star, the first-magnitude Achernar, meaning "end of the river" in Arabic. This star is the ninth brightest in the sky.

The winding constellation has a few interesting stars. Omicron-2 (s2) is a multiple star, which small telescopes will readily resolve into a pair. In larger telescopes, the fainter of the two reveals itself to be a pair of dwarfs, one red, the other white. Further west lies Epsilon (e), which is the third-nearest naked-eye star to us, at a distance of 10.7 light-years. It is also the nearest Sunlike star.

FORNAX, THE FURNACE

A modern constellation (1750s), the Furnace commemorates the advanced iron-smelters that were helping to drive the Industrial Revolution. It is not a bright constellation, and is of interest mainly to telescopic observers. They will be able to spot southwest of Beta (β), a small irregular galaxy called the Fornax System, which is a member of our Local Group of galaxies. Southeast of Beta, near the triangle of stars Phi (φ), Gamma (γ) and Eta (η) in neighboring Eridanus, is the Fornax cluster of galaxies. At a distance of over 60 million light years, it is one of the nearest clusters to us.

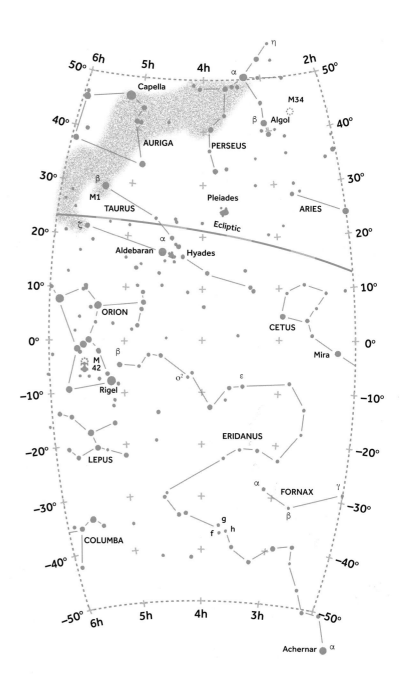

▲ Constellations visible near the meridian at about 11pm during the first week in December.

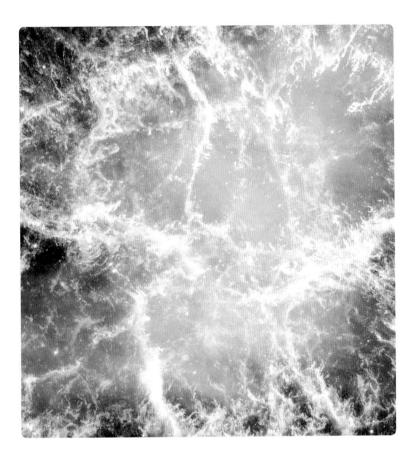

▲ Above: The Crab Nebula in Taurus was born in a supernova, spotted nearly 1,000 years ago.

▲ Above right: Nebulosity surrounds the stars in the Pleiades cluster. They are white, hot, and young.

TAURUS, THE BULL

One of the easiest constellations of the Zodiac to spot is Taurus, visualized as a charging bull, with a baleful red eye and horns. The Sun travels through the constellation between May 14 and June 21 every year.

In mythology, Taurus is the bull that features in one of the many seductions of Zeus. To capture the attention of the lovely Princess Europa, he turned himself into a handsome white bull. She climbed on his back, and he swam to the island of Crete, where he resumed his manly form. One of their children became King Minos, who built the famous labyrinthine palace at Knossos and encouraged bull worship.

Highlights abound in Taurus. Aldebaran is the noticeably reddish eye of the Bull. It is surrounded, in sight but not in space, by the open star cluster called the Hyades. This cluster has a prominent "V" shape and is easily seen with the naked eye, with its brightest stars being of about the fourth magnitude. At a distance of some 130 light-years, the Hyades is one of the nearest open clusters (but it is twice as far away as Aldebaran).

Farther north is the open cluster of the Pleiades (M45). It is commonly called the Seven Sisters, but exceptional eyesight is needed to make out its seven brightest stars. Alcyone (magnitude 2.9) is the brightest. With binoculars, many more stars show up in the cluster, which is estimated to contain well over 500 stars in all. They are relative youngsters, around 10 to 20 million years old. The center of the Pleiades cluster lies about 450 light-years away, much farther than the older Hyades.

Another highlight of the constellation was the first object in Messier's Catalog, M1, more popularly known as the Crab Nebula. It is what remains of a supernova explosion that took place in July, AD 1054, and was witnessed by Chinese astronomers. They recorded that this "guest star" shone so brightly that it could be seen in the daylight for more than three weeks. The supernova remnant is still expanding, but is only visible with a telescope.

PERSEUS

Perseus was one of the great heroes of Greek mythology. It was he who beheaded the dreaded Medusa, the Gorgon who had serpents for hair and whose gaze would turn mortals into stone. And it was he who, heading home after this adventure, rescued Andromeda from the jaws of the sea monster Cetus.

This far northern constellation has much to commend it. The asterism formed by Beta (β) and its neighbors represents the head of Medusa. Beta is famed as the "Winking Demon" Algol. It is a variable star of the eclipsing binary type, in which a bright and a dim star periodically eclipse each other. Beta's brightness, therefore, dips, making it appear to "wink." This happens as regularly as clockwork every 2 days, 21 hours, when it dims from the second to the third magnitude for about 9.6 hours. Algol was the first eclipsing binary to be recognized, by the English astronomer John Goodricke in 1782. Northwest of Algol and about halfway to Gamma (γ), in neighboring Andromeda, is an open cluster (M34) that is just visible to the naked eye.

December Stars

TOP: LOOKING SOUTH

Constellations visible in North America and Europe at about 11pm on or near December 7

To the west of the meridian, the skies are relatively dull, dominated by Cetus and Eridanus, as well as Pisces, Andromeda, and Pegasus. To the east of the meridian, Orion is in mid-skies, with Taurus higher up and Canis Major lower down. Canis Minor and Gemini are also well above the horizon. Sirius, Rigel, Aldebaran, Capella, Pollux, and Procyon form a polygon of bright stars.

BOTTOM: LOOKING NORTH

Constellations visible in North America and Europe at about 11pm on or near December 7

Cygnus has not yet reached the western horizon, but Aquila has now disappeared beneath it. Vega is still visible low in the east, but the Summer Triangle is no more. The handle stars of the Little Dipper are now on the meridian. In the far west, the Square of Pegasus is prominent, but not for long. In the far east, Leo is rising, with bright Regulus and the curve of stars above it making the familiar sickle shape.

NORTHERN HEMISPHERE

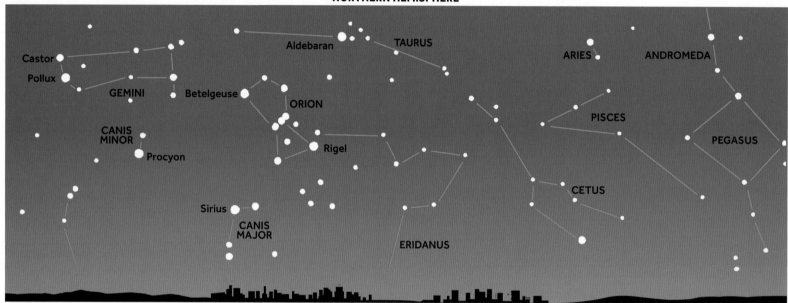

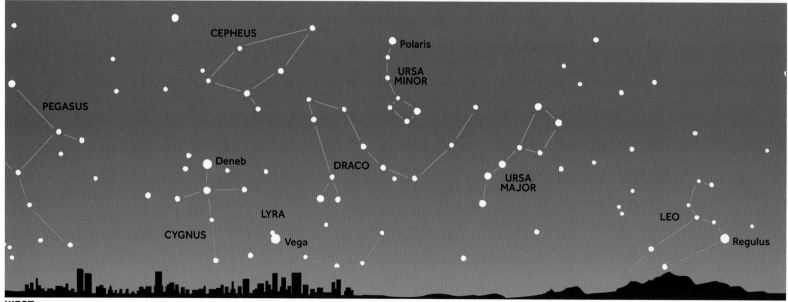

WEST   NORTH   EAST

TOP: LOOKING SOUTH

Constellations visible in Australia and South Africa at about 11pm on or near December 7

The Southern Triangle (Triangulum Australe) is back on the meridian this month, but low down. Six months ago, in June, it was also on the meridian, but high up. This reminds us that the heavens have turned half-circle since then. Most of the interest this month lies in the southeast, where the brilliant constellations now lie. West of the meridian the skies are dull, with only Achernar and Fomalhaut shining out.

BOTTOM: LOOKING NORTH

Constellations visible in Australia and South Africa at about 11pm on or near December 7

To the west of the meridian, the skies are bare, dominated by faint constellations like Cetus and Pisces. East of the meridian, Taurus is in mid-skies, with Orion higher still and Canis Major further east. Canis Minor, Gemini, and Auriga are also well above the horizon. Sirius, Rigel, Aldebaran, Capella, Pollux, and Procyon are also visible together in the sky.

SOUTHERN HEMISPHERE

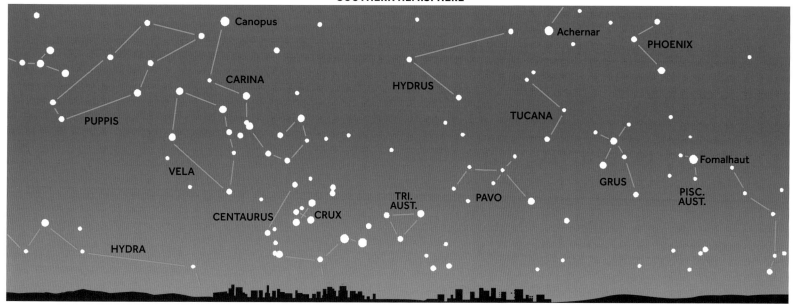

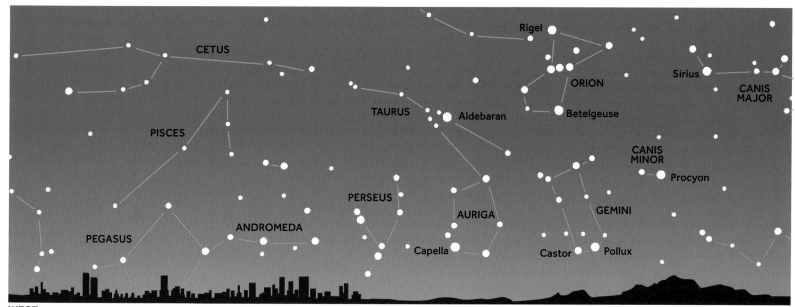

WEST   NORTH   EAST

Milestones in Astronomy

3000s BC	Babylonian and Egyptian astronomy develops in the Middle East.
2000 BC	Ancient Britons begin to build Stonehenge, which will become one of the first astronomical observatories.
585 BC	The Greek philosopher Thales of Miletos predicts this year's solar eclipse.
240 BC	First definite sighting of what was to become known as Halley's comet.
200s BC	Aristarchus of Samos suggests that Earth moves around the Sun. Eratosthenes measures the circumference of Earth with surprising accuracy.
150s BC	Hipparchus compiles a catalog of more than 1,000 stars. He discovers precession.
AD 150	Ptolemy of Alexandria writes a seminal astronomical treatise detailing existing knowledge, including the concept of an Earth-centered universe—the Ptolemaic Universe. It becomes known in its Arabic translation as *Almagest*, "The Greatest."
1054	Chinese astronomers witness the supernova explosion that resulted in the Crab Nebula.
1543	Nicolaus Copernicus proposes a Sun-centered universe, contrary to the Church's belief in Ptolemy's Earth-centered view.
1572	Supernova witnessed by Johannes Kepler, and becomes known as Kepler's star.
1600	Giordano Bruno is burned at the stake for supporting Copernicus's theory.
1604	Supernova witnessed by Tycho Brahe, now called Tycho's star.
1609	Galileo makes the first telescopic observations of the heavens. Johannes Kepler publishes his first laws of planetary motion, one stating that the planets move around the Sun in elliptical orbits.
1666	Isaac Newton formulates his laws of gravity.
1668	Newton builds the first reflecting telescope.
1675	Greenwich Observatory founded in London.
1705	Edmond Halley predicts (correctly) the return (in 1758) of the comet now named after him.
1781	William Herschel discovers Uranus.
1783	John Goodricke explains the light variation of Algol, the first eclipsing binary to be recognized.
1784	Charles Messier publishes a catalog of nebulae and star clusters.
1801	Giuseppe Piazzi discovers Ceres, the first asteroid.
1802	William Herschel introduces the term "binary star."
1838	Friedrich Bessel measures the distance to a star, 61 Cygni, using the method of parallax.
1845	The Earl of Rosse discovers the spiral nature of a nebula, later to be identified with the Whirlpool Galaxy M51.
1846	John Adams and Urbain Le Verrier independently work out the position of a new planet, which Johann Galle then discovers—the planet Neptune.
1877	Giovanni Schiaparelli reports he has seen "canali" on Mars, triggering the idea of a race of intelligent Martians.
1882	This year's comet is one of the most brilliant ever recorded.
1888	John Dreyer publishes his *New General Catalogue of Nebulae and Clusters of Stars*—NGC numbers are still widely used.
1901	A brilliant nova is visible in Perseus.
1905	Albert Einstein announces his general theory of relativity, which revolutionizes ideas of space, time, and motion.
1910	Return of Halley's comet, which was outshone by the "Daylight comet."
1912	Henrietta Leavitt discovers the period-luminosity relationship of Cepheid variable stars. Vesto Slipher discovers the red shift of extragalactic nebulae.
1913	Henry Russell discovers the relationship between stellar luminosity and spectral type, simultaneously with Ejnar Hertzsprung, and they establish the Hertzsprung-Russell (H-R) diagram.
1918	100-inch (2.5-m) Hooker telescope comes into use at Mount Wilson Observatory near Pasadena, California.
1925	Edwin Hubble discovers galaxies beyond our own using the Hooker telescope.
1930	Clyde Tombaugh discovers Pluto.

1931	Karl Jansky discovers radio waves coming from the heavens and pioneers radio astronomy.	**1997**	Comet Hale-Bopp becomes one of the brightest comets of the century.

1931 Karl Jansky discovers radio waves coming from the heavens and pioneers radio astronomy.

1948 200-inch (5-m) Hale telescope completed at Mount Palomar Observatory near Pasadena, California.

1955 Jodrell Bank radio telescope becomes operational in Cheshire, England.

1957 First artificial satellite, *Sputnik 1*, launched.

1958 First U.S. satellite, *Explorer 1*, discovers Earth's Van Allen radiation belts.

1959 Russia's *Luna 3* probe photographs the Moon's far side.

1960 Quasars are discovered.

1962 First successful planetary probe, *Mariner 2*, reports on Venus.

1965 *Mariner 4* takes the first close-up pictures of Mars. Comet Ikeya-Seki is splendid seen with the naked-eye. Arno Penzias and Robert Wilson pick up background microwave radiation of 3K (3 degrees above absolute zero), in line with the Big Bang theory.

1967 Astronomers in Cambridge, England, discover pulsars.

1969 *Apollo 11* astronauts land on the Moon, on July 20. Neil Armstrong plants the first footprints in the lunar soil.

1974 *Pioneer 10* investigates Jupiter.

1976 *Viking 1* and 2 probes land on Mars.

1978 Rings are discovered around Uranus.

1979 *Voyager 1* and 2 probes fly past Jupiter. *Pioneer 11* reports from Saturn.

1981 The first space shuttle is launched.

1986 *Voyager 2* flies by Uranus. Halley's Comet returns, pictured close up by Europe's *Giotto* probe.

1987 Supernova visible with the naked eye spotted in the Large Magellanic Cloud, designated 1087A.

1989 *Voyager 2* flies by Neptune.

1990 Hubble Space Telescope launched from space shuttle.

1994 Fragments of Comet Shoemaker-Levy 9 crash into Jupiter.

1996 Comet Hyakutake easily visible to the naked eye.

1997 Comet Hale-Bopp becomes one of the brightest comets of the century.

1998 *Lunar Prospector* discovers deposits of ice on the Moon.

1999 Total eclipse of the Sun on August 11 passes through some of the most populous regions of the world from England to India. Most European sightings spoiled by cloudy weather.

2001 *Near Earth Asteroid Rendezvous (NEAR) Shoemaker* lands on Eros. This is the first soft landing on an asteroid.

2003 China launches an astronaut into space, becoming the third country in the world to have a human space program.

2004 The *Huygens* lander finds lakes of liquid hydrocarbons on Titan, the largest moon of Saturn.

2006 Pluto and Ceres are reclassified as dwarf planets along with several large bodies found in the Kuiper Belt.

2011 The *Messenger* probe becomes the first to go into orbit around Mercury.

2012 A survey by the Kepler space observatory estimates that the number of planets in the universe is higher than the number of stars.

2015 After nine years in space, *New Horizons* flies past Pluto, revealing its icy surface for the first time.

2016 The first observations of gravitational waves are made. Gravitational waves are oscillations in spacetime caused by the movement of massive objects. In this case the waves were caused by a pair of colliding black holes.

2017 Oumuamua, a 755-ft (230-m), cigar-shaped rock, is observed passing the Sun. Its trajectory shows that it originated outside the solar system and is therefore the first interstellar object to be observed. It is traveling so fast it will leave the solar system in 20,000 years.

2018 Observations reveal that the galactic core of the Milky Way contains dozens, possibly as many as 10,000, black holes.

Index

Photo Credits

About the Authors

Robin Kerrod is the author of *Discover the Night Sky*, published in 2000, which has been thoroughly edited and updated by Tom Jackson to include new events and developments in the field for this book.

Robin Kerrod is a Fellow of the Royal Astronomical Society and the British Interplanetary Society, and is the author of many bestselling books on astronomy, space, and the sciences. He lives near Salisbury, England.

Tom Jackson is a science writer based in Bristol, UK. He has written numerous books covering all kinds of subjects from axolotls to zoroastrians. He has worked on several astronomy books, including titles with Brian May and Patrick Moore.